The 21st Century Fight for the Amazon

Mark Ungar
Editor

The 21st Century Fight for the Amazon

Environmental Enforcement in the World's Biggest Rainforest

palgrave
macmillan

Editor
Mark Ungar
Brooklyn College and Graduate
Center
City University of New York
New York, NY
USA

ISBN 978-3-319-56551-4 ISBN 978-3-319-56552-1 (eBook)
https://doi.org/10.1007/978-3-319-56552-1

Library of Congress Control Number: 2017947809

Cover illustration: Pattern adapted from an Indian cotton print produced in the 19th century

Printed on acid-free paper

This Palgrave Macmillan imprint is published by Springer Nature
The registered company is Springer International Publishing AG
The registered company address is: Gewerbestrasse 11, 6330 Cham, Switzerland

Acknowledgements

When I first told people about this project on the Amazon, many thought it was about the online shopping giant. Even amid growing alarm at climate change, it was a reminder of how disconnected we have become from its root causes. So I am indebted to all the organizations and people contributing to this book's effort to understand those causes. For its initial funding and confidence, I would like to thank the Inter-American Development Bank. During fieldwork in the Amazon Basin, many organizations unhesitatingly lent their deep expertise and scarce time. They include INPA (*Instituto Nacional de Pesquisas de Amazônia*) and the Center of Integration and Improvement in Environmental Policing (CIAPA: *Centro de Integração E Aperfeiçoamento em Policía Ambiental*) in Brazil; Ecuador's Green Vigilance Unit (*Vigilancia Verde*), the Bolivian Forest and Land Authority (*Autoridad de Bosques y Tierra*); and most of all the environmental courts, prosecutor's offices, and police units in every country. This project would also have been impossible without the dozens of non-governmental organizations who shared their wealth of knowledge. They include groups affiliated with the book's authors, such as the Federal Police of Brazil, the Amazon Conservation Team, OEFA, Peru's Proyecto Amazonas, and Janette Ulloa of Pro-Ideas. During meetings and on patrols, I guiltily heard logged trees crashing through windows of opportunity opened by their distracted attention toward this book. Their courage, hard work, and sheer stamina are why the fight for the Amazon has a chance. Chasing an AWOL llama down a highway in La Paz with the SPCA or crossing a washed-out bridge with investigators in Venezuela were adventures for a privileged North American but a fraction of the innumerable hardships faced every day by Latin American environmentalists.

For their meticulous translations, I would like to thank Judith Filc, Alix Shand, and Maria Antonieta Merino de Taboada. Most of all, I would like to thank my husband, Bob, for the values and love that keep me on track, and our son, Dylan, whose enthusiasm on one trip to the Amazon let me see the forest through the eyes of the future.

Contents

Contributors

Manuel Joel Díaz Capdevilla is President and Editor of the first Ecological Channel of Venezuela (*Primer Canal Ecológico de Venezuela, ECOVISION*); Secretary General, Ecological Movement of Venezuela; and Coordinator, National Congress of Green Citizenry (*Congreso Mundial de Ciudadanía Verde*).

Luz Marina Mantilla Cárdenas is Director General of the Amazon Institute of Scientific Research, SINCHI (*Instituto Amazónico de Investigaciones Científicas*).

María Antonieta Merino de Taboada is Former Head of the Oversight Body of the Agency for Environmental Assessment and Enforcement (OEFA) of Peru.

Katia Delvoye works with the Amazon Conservation Team (ACT), Suriname's leading Amazon protection organization.

Hugo R. Gómez Apac is Former Chairman of the Board of the Agency for Environmental Assessment and Enforcement (OEFA), Peru's top national environmental enforcement body.

Hermes Libretto works with the Amazon Conservation Team (ACT), Suriname's leading Amazon protection organization.

Víctor López Acevedo is a professor and co-founder of Corporación Pro-Ideas, an Ecuadorian socio-environmental studies institute that conducts collaborative studies with local producers and officials of the Amazon.

Milagros Granados Mandujano is Former Technical Secretary of the Environmental Court in Peru.

Minu Parahoe is the Program Director of the Amazon Conservation Team (ACT), Suriname's leading Amazon protection organization.

Franco Perazzoni is the Regional Chief Officer for Investigations and Organized Crime (DRCOR) of the State of Amazonas of Brazil, where he directs all environment and organized crime investigations. He was former head of environmental crimes unit, State of Mato Grosso (2005–10) and Director, Central Federal Unit, Environmental crimes (2010–14).

Ana María Hernández Salgar is Director of International Affairs, Politics, and Cooperation of the Alexander von Humboldt Institute of Biological Reseach (Instituto de Investigación de Recursos Biológicos) of Colombia.

Manlio Alberto Roca Zamora is former Minister of Sustainable Development and the Environment, Department of Santa Cruz, Republic of Bolivia.

Mark Ungar is Professor of Political Science and Criminal Justice, City University of New York.

List of Figures

List of Tables

CHAPTER 1

Introduction: The Evolution of Environmental Enforcement

Mark Ungar

Abstract This book's introduction describes the emerging structure of environmental enforcement, which the Amazonia has pioneered for the world, centered on the formation of police agencies, prosecutor units, and special courts. It shows how this structure, by giving the environment effective political and institutional support, is a historical and legal breakthrough in global protection. But the book's analysis and case studies will also explain how that same structure is weakened by the larger political, institutional, and geographic context in which it operates.

Keywords Amazon · Climate change · Democracy · Environment Latin America

This book brings together an extraordinary set of specialists who are tackling the extraordinarily complex threats to the Amazon Basin, the home to a full half of the world's remaining tropical forests and well over

M. Ungar (✉)
Brooklyn College and Graduate Center, City University of New York, New York, NY, USA
e-mail: MUngar@brooklyn.cuny.edu

M. Ungar (ed.), *The 21st Century Fight for the Amazon*,
https://doi.org/10.1007/978-3-319-56552-1_1

half of its plants and animal species. If the Amazon was a nation, it would be the world's 9th biggest. With five of just 17 countries worldwide considered "mega-diverse," it produces more than 20% of the planet's oxygen and stores some 120 billion tons of carbon each year—over 17 times the amount the entire US spews out. But the Amazon has also been a perennial treasure trove. From the search for El Dorado in the 1500s to the rubber barons of the 1800s to the ranchers of the 1900s, the Amazon has been coveted as a source of boundless riches and subject to an endless bloodletting of exploitation.

Only recently has the balance between pillage and protection finally shifted. Building on global environmental movements and programs like the 2005 United Nations-affiliated program Reducing emissions from deforestation and forest degradation (REDD+), people throughout the Amazon assiduously and courageously forged innovative initiatives of legal enforcement and sustainable development—many of which this book's contributors helped pioneer—that have become models of conservation worldwide. They range from local programs like green municipalities and community policing to larger schemes of certified timber and protected areas. And they are all boosted by technological leaps like satellite monitoring, which provides data that is skillfully used in policy and public campaigns by organizations like the Amazonian Network of Geo referenced Socio-Environmental Information (RAISG), which includes many of this book's authors. As a result of all these efforts, deforestation in the Amazon began to decline for the first time since rates started being tracked. In 2004 alone, an estimated 2.8 m hectares (10,700 square miles) of the rainforest were razed — an area larger than the state of Massachusetts. But in 2010, that rate dropped by over 66% to about 750,000 hectares (3000 square miles). Deforestation fell in nearly every country; in Colombia and Peru, the annual rate from 2010 to 2013 was half of the rates of the 2005–2010 period.

But ecocide is always one step ahead. While deforestation hit an all-time low in 2011, it is poised to rebound. In Brazil, it rose 28% in 2012 and 2013; nearly two million acres were cut down in that country from August 2015 to July 2016, a sharp jump from the 1.5 million recorded between August 2014 and July 2015. In Venezuela, the rate of 2010–2013 was double the rate of 2005–2010. And in every one of the Basin's countries, the amount of territory designated as concessions for corporate extraction of hydrocarbons and minerals, or which is being inundated by small-time miners, is at a record high. So too is the amount of land being cultivated with unsustainable exports crops like palm oil and cacao. The myth of the Amazon's untapped wealth remains as seductive and destructive as it is ultimately elusive. Rather than being emulated, protected areas have become besieged. Rather than being subsidized, sustainable economies have been displaced. Rather than being coveted as greenhouse gas absorbers, 36–57% of the region's 15,000 tree species are under threat of extinction. The impressive but shaky edifices of legal protection are easily toppled by the governments that built them or eroded by countless acts of fortune-seeker desecration underneath them. So even as Latin America's environmental laws become clearer and more comprehensive, they are also becoming increasingly incapable of halting deforestation, mining, incursions into protected areas, illegal road building, land appropriations, and other ruinous practices.

This book helps explain why they struggle to halt this onslaught. In its wealth of policy analysis and case studies, it describes a region caught in the pincers of fusion and fragmentation. From one side is a regional move toward extraction of oil and minerals for a ravenous global economy. This race has erased ideological boundaries, bringing in regimes spanning the political spectrum, from the neoliberal right to the socialist left—as amply discussed in the chapters on Venezuela, Bolivia, and Ecuador, which are led by the region's most prominent leftists. The commodities boom is simply too tempting; the paradox of plenty too conveniently ignored. Even decades of sustainable development policy—from rubber tapping and medicines to eco-tourism—are still no match for grandiose mass extraction. There is also a consolidation of organized crime, whose power burns with the perpetual fuel of high prices, corrupt officials, and opportunities for collaboration. From the other side comes fragmentation. The establishment of new enforcement agencies is a key advance, but counter-productive if they lack support or are corrupt. As discussed in several chapters, in addition, Latin America's unprecedented decentralization has expanded governance but also multiplied the agencies unprepared to resist fraud and coordinate with each other. The criminal justice system too is frayed, with weak links—such as a lack of prosecutors with environmental training—that do not just slow down an already sluggish investigatory process, but often derail it. Protected areas and indigenous peoples are also under siege by a swarm of small-scale miners and large-scale fortune-seekers. This book describes how these threats emerge and how they can be contained and possibly reversed.

The Advent of Environmental Enforcement: Ten Advances

Environmental destruction has become so difficult to curb in part because it is so wide-ranging. Even defining it is a challenge. In the simplest typology, most of it falls in two broad categories: illegal removal—of flora, fauna, minerals, and other natural resources; and illegal addition—of elements, from gold-mining mercury to construction waste. Controlling such a wide spectrum of actions requires an even wider set of enforcing tools to cover each one at each stage, from prevention to prosecution. Among them are regular water testing, protection of witnesses, physically stopping unauthorized transport, monitoring sawmills, deploying nature reserve guards, dismantling drug cartels, and promoting community policing—a tall order for countries still documenting the problem or willfully ignoring it. But ten advances, which helped bring down deforestation after 2004, are a foundation for future progress.

1. *Constitutional and Penal Law:* The countries of the Amazon are among the first not just to include the environment in their constitutions, the highest level of law, but give them full constitutional protection. Around the world, just 24 countries include the full range of environmental protections; of them, five are in the Amazon.[1] In 2008, Ecuador was the first country to constitutionally recognize nature's rights—on a level par with humans. Such protection has been bolstered with stronger law in the region. Brazil's 1998 Environmental Crimes Act is regarded as one of "the most modern and comprehensive legal texts focus on environmental crime" (Division of Environmental Law and Conventions 2012), while Venezuela's 2012 Environmental Penal Law gives the statewide leeway against environmental crime. Bolivia created a legal framework for Mother Earth, or Pachamama, while Ecuador added more specific environmental crimes (article 246) to its penal code in 2014.
2. *Environmental Ministries:* Cabinet-level environmental ministries give the environment political power and protection. Nearly every Latin American country has established such a ministry, some of which were separated out from larger ministries, in order to better provide regulations, funds, and policy development. As discussed in the scholarship on Latin America, institutional strength within

the state is key to policy effectiveness, and a ministry is a keystone of that political and policy framework. Connected to and encouraged by these ministries have been new units formed to consolidate and direct enforcement efforts. One of the most prominent units was established by Peru, whose directors wrote this book's chapter with a detailed focus on how it methodically documents and fights environmental crime in its vast Amazon region.

3. *Environmental Police:* The police long have been empowered to enforce environmental law, but a poor record and the distinct nature of that task has spurred formation of special environment units within the security structure. The need for evidence and witnesses that only on-the-ground policing can attain, along with successes in other regions against animal trafficking and waste exports, have also encouraged this approach. In the global South, as Table 1.1 shows, Latin America pioneered this form of policing, with environmental police units in nearly half of its countries. Amid decentralization, provinces and municipalities are also forming local bodies. And as the chapter on Brazil shows, the need for policing is as great as its potential. Written by one of the country's top law enforcement officials, it describes how the federal police turns statistics into firepower—using combinations of electronic surveillance, satellite imagery, and institutional coordinations to swoop down red-handed on environmental crime.
4. *Prosecutors:* Since environmental law means little without bringing justice to violators, special new units within the Attorney General's Office (MP: *Ministerio Público*) have been established in 13 Latin American countries. Many of them train prosecutors (*fiscales*) in environmental law and establish local offices around the country. The authorities and responsibilities given to these units vary widely. Prosecutors in some countries can promote public mobilization against environmental crime, for example, while other countries allow local branches to negotiate non-penal resolutions for violations. In all cases, though, they are backed up by the state's power to prosecute the full range of crimes stipulated in increasingly detailed environmental law.
5. *Environmental Courts:* A third pillar of the state's environmental enforcement structure is environmental courts to rule on specific cases and adjudicate environmental law in general. As of 2015, six Latin American countries—Chile, Costa Rica, Brazil, Mexico,

Table 1.1 Primary environmental enforcement agencies in Latin America, by Country, 2015

Country	*Police*	*Criminal Justice Units*
Argentina		Unidad Fiscal de Investigación en Material Ambiental
Bolivia	Policía Forestal y de Medio Ambiente	
Brazil	Unidad do Policía Federal	Ministerio Público
Chile	Jefatura Nacional de Delitos Contra Medioambiente y Patrimonio Cultural	
Colombia	Policía de Medio Ambiente	Unidad Nacional de Fiscalías de Delitos contra los Recursos Naturales y el Medio Ambiente
Costa Rica		Fiscalía Adjunta Agrario Ambiental
Dominican Republic		Procuraduría para la Defensa del Medio Ambiente
Ecuador	Unidad de Protección de Ambiente	Fiscales de Medio Ambiente
El Salvador	División de Medio Ambiente	Unidad de Medio Ambiente de la Fiscalía General
Guatemala	Unidad Ambiental, Ministerio de Seguridad	Fiscalía de Delitos Contra el Ambiente
Honduras		Fiscal Especial del Medio Ambiente
México		Fiscalía Especializada para la Atención de Delitos Ambientales
Nicaragua		Fiscalía General de la República
Panamá		División de Delitos Ambientales, Dirección de Investigaciones Judiciales
Paraguay		Unidad Fiscal de Delitos Ambientales
Perú	Policía del Medio Ambiente	Fiscales Especiales del Medio Ambiente
Uruguay		
Venezuela		Dirección General de Medio Ambiente

Source Created by author

Peru, and Bolivia—have separate courts or court branches for environmental cases. In a positive trend, more recently created ones have more jurisdictional authority and specialized structures, key to the state power discussed below. For example, in December 2014, Peru established high-level courts for mining, energy, and fisheries and manufacturing.

Table 1.1 summarizes the primary agencies of environmental enforcement in Latin America, highlighting their greater extent in the Amazon Basin countries.

6. *Satellites:* GPS technology has taken enforcement to a new level by providing real time and universal coverage of the Amazon. Large-scale clearing is now far more easily detected and curbed—in fact, many studies estimate that such action has led to more than three-fifths of the observed drop in deforestation since 2004. Each Amazon country is incorporating such technology; for example, Peru has adopted a program called GeoBosques. But Brazil has applied this approach most aggressively, using its satellite systems, DETER (Deforestation in Real Time) and the Amazon Protection System (SIPAM: *Sistema de Proteção da Amazônia*), to channel intelligence from bi-weekly observance to command-and-control police operations against larger scale forest clearing. Satellite statistics have also been taken up by the region's science and research organizations, such as Brazil's National Institute of Amazon Research (INPA: *Instituto Nacional de Pesquisas da Amazônia*) and the regional RAISG. And from all this big data comes criminal evidence—a lifeline for beleaguered enforcement and judicial officials on the ground.
7. *Non-Governmental Organizations (NGOs) and Social Movements:* As with the human rights and environmental movements of which they are a part, the flourishing of citizen-based organizing in the Amazon is perhaps the single most effective and multipurpose catalyst for vigorous enforcement. The wide range of NGOs—from biology research institutes to neighborhood seminars—provide detailed information, monitor extraction sites, and educate citizens. They have also helped pave the way for the region's fast-growing grassroots social movement. Inspired by the advent of Third Generation Human Rights, which include the right to a healthy environment, this movement ranges from cross-border networking, such as the nascent anti-mining movement in Latin America, to confrontation with the state, such as the 2009 protest

by indigenous communities in the Amazon town of Bagua in Peru.

8. *Protected Areas:* The clearest demarcation of environmental protection is physical. Throughout the region, an unprecedented amount of land has been demarcated for protection against economic activity from the outside and for the preservation of indigenous peoples within. In Peru, for example, 74% of the deforested Amazon areas from 2000 to 2015 occurred outside protected natural and indigenous areas, while deforestation rates in non-protected indigenous areas rose 68% between 2011 and 2015. Many of this book's authors stress the importance of such areas. In the chapter on Ecuador, one of the country's leading scientists delves into how that country's elaborate system of protected areas is both managed and mismanaged.
9. *Non-State Initiatives:* Following the policy trend in global climate change, as well as the state's loosening economic grip, is a shift from command-and-control (CAC) policies toward market-based initiatives (MBIs). MBIs are now at the forefront of environmental enforcement because they are regarded as able to create the most effective and self-enforcing forms of environmental protection that minimize costs and corruption. Many EIs are macro-economic, such as fiscal or budgetary measures like biodiversity allocations and discharge fees. Most innovative ones are micro-economic, though, such as sustainable goods promotion and revenue-sharing arrangements. Such programs have promoted markets for a range of goods, from certified timber and fair-trade coffee to quintessentially Amazon products like Acai and camu camu. In this book's chapter on Bolivia, the former environmental chief of Santa Cruz province, the country's agro-industrial workhouse, describes how it has encouraged such initiatives.
10. *Municipal Programs:* Spreading throughout the Amazon are local initiatives. One of the most common are Green Municipalities. Adopting the benchmarks now in vogue for local governance, city administrators are creating goals for re-forestation, adoption of better agricultural practices, forest areas under management plans, smallholder settlements, and low-carbon energy revenues. In some cases, such as in the Brazilian states of Pará and Tocantins, where many of the initial green municipalities first sprung up, they are like seedlings of a landscape stripped bare by mining and

deforestation. Other programs extend deep into rural areas. For example, one initiative in the Peruvian Amazon established an office to file complaints on behalf indigenous communities that have long been exploited by loggers and other businesses.

Ten Challenges

Recognition of these advances, though, often miss the forest for the trees. They are far-reaching and diverse, together forming a framework for action. The real test for these efforts, though, comes when they invariably conflict with other parts of the state, organized crime, many parts of society, the private sector, and, in many cases, against each other. Raids are satisfying, but real progress comes in the long and sluggish criminal justice process, from investigation to conviction. Crisp laws and determined organizations are awarded with praise but are built on a bog of weak states and double-crossing politicians. And the double sword of technology enables not just detection of illegal extraction, but new ways to make it wider, deeper, and more justified by its smaller carbon footprint.

1. *Fragmented States:* As it expands, the state's environment structure risks coming apart. As discussed throughout the book, coordination among agencies—within the environmental network and with other agencies—is complicated at best. With more agencies comes more institutional fratricide, political grandstanding, and, for the most unwary, the stealth attacks of dwindling budgets. More officials also means more corruption, making enforcement by isolated officers in the Amazon—as demonstrated through this book—more akin to exploitation. Such difficulties are deepened by the unprecedented level of decentralization throughout Latin America since democratization began in the 1980s. In the chapter on Suriname, the country's leading environmental organization describes how a disorganized state administration opens the door for destruction of its extensive rainforests. In Bolivia, even more dramatically, the decentralization that started in 2006 has led to hundreds of new local governments, thousands of local officials, and a corresponding increase in local environmental agencies such as Committees of Environmental Monitoring and Municipal Forest Units.

2. *Geography:* Matching and mirroring the state's fragmented nature is that of the lands it governs. A central theme in Latin American studies is the greatly varying levels of legitimacy, presence, and effectiveness by state agencies in a country's territory (see O'Donnell 1993). Few issues demonstrate such variation as clearly as the environment, which extends from urban sanitation systems to remote rainforest groves. But even with its growing structure, the state's reach is limited. One of the most arresting depictions of its truncated control are the Amazon's ever-extending roads. On one level, the highway built by the state are a main cause of the deforestation is now fights. About 85% of Amazon deforestation is concentrated up to 50 km from the main roads (Chomitz and Thomas 2001). Those routes then open the gateway to even greater deforestation by spurring extensive "fishbone" patterns of secondary roads that fan out from the main routes in every country and provide access to the full range of environmental crime. In Brazil alone, over 105,000 miles of such roads are unauthorized.
3. *Investigation:* Along with the many existing limits of general law enforcement in Latin America, investigation of environmental crime is hindered by obstacles such as physically remote locations; the rapid contamination or degradation of physical evidence; and poor coordination among police and scientists. Such limits are amplified at the local level, where there is inadequate expertise even on daily issues like solid waste. In the rural areas where most environmental crime occurs, this lack of staff and infrastructure then means dispatching city-based specialists, who usually don't arrive on time to secure the witnesses or proof sufficient for trial—a stark illustration of green criminology's critique, discussed below. For operations in Lábrea, the southernmost municipality of Brazil's Amazonas State, officials often need to trek overland through Acre and Rondônia states. Back in the cities, investigation is further hobbled by the lack of experts, laboratories, and prosecutors (*fiscales*). In Ecuador, for example, the only *fiscal* with advanced environmental training is assigned to too many other issues to be able to concentrate on them. "Judges and *fiscales* do not [necessarily] know what constitutes a crime and do not know how to prosecute them", that *fiscal* says. Although four articles of Bolivia's Law 1333 mention use of the Penal Code to prosecute crimes, there are no prosecutors or viable procedures

to do so. As a result, there have been "very few interventions in which the government has sanctioned businesses that have caused environmental damage."[2] In Colombia, one of the few *fiscales* with an environmental portfolio says that the *Fiscalía* is inundated with up to a hundred complaints per day, some running to thousands of pages. Prosecutors add that police and judges can take up to six months to take steps as basic as decommissioning caches of seized logs. Stuck in legal limbo, such contraband often generates lawsuits by its aggrieved owners—tying up cases for even longer. Investigation is hampered by state indifference as well as inefficiency. In Ecuador, officers of the police's environmental branch, the *Unidad de Protección de Ambiente* (UPMA), complain that serious cases like arson in eco-sensitive lands are met by "a lack of attention by *fiscales* and judges." Compounding the state's indifference is its complicity. In Venezuela, Amazonas state is being decimated by illegal miners who enjoy the protection of the armed forces, which are responsible for environmental protection.[3] In neighboring Colombia, state forest managers are routinely pressured or bribed into falsifying records of logs extracted in protected areas. In Bolivia, where environmental police officers have charged up to US$1700 from each driver transporting illegal wood, an elaborate network of corruption was exposed in 2012.[4] "The State knows who traffics wood. Those with the backing of the government" often unrestricted trafficking, even across international border said one official.

4. *Organized Crime:* Investigation is further hindered by the increasingly organized nature of environmental crime. Worldwide, the value of environmental crime is estimated to be between $91 and 258 billion today, sharply up from the 2014 estimate of $70–213 billion, according to the two main international enforcement agencies.[5] One reason for the growth is that, in regions like the Amazon, criminal networks use the environment as a cover, opportunity, and gateway. Not only are goods such as timber and animals are highly lucrative—some ornamental fish cost two dollars on the local black market, for example, but well over a hundred abroad. But characteristics of the areas from which they are removed—like powerful local families and weak state agencies—are ideal for hiding and moving them. They also offer abundant opportunities for collaboration among environmental and

organized crime, such as when loggers clear the way for drug syndicates or provide arms for local groups, such as the Amazon's many private militias, or even state-based entities like some "indigenous police" units in Brazil.[6] The horrific prison massacres in prisons throughout the Brazilian Amazon in January 2017, in fact, were organized by the Família do Norte, which controls drug trafficking in the region. Organized crime networks are also skillful at evading the law through displacement, which can be spatial (to another area), temporal (to a later time), or transitional (another kind of offense). Officials in Brazil and Ecuador, for example, complain about illegal networks they break up later reconvening as "eco-tours" that are fronts for illegal hunting and poaching. Organized crime "succeeds by identifying, re-enforcing and exploiting links and weaknesses in the state," asserts a Bolivian judge who adds that illegal settlements in protected areas are bankrolled by Colombian and Mexican cocaine cartels.

Such conditions draw unspoken peripheries of criminal investigation throughout Latin America. With an often self-fulfilling belief that they are doomed to chase symptoms instead of causes, most units hesitate to take action that might get them tangled up in organized crime. Officers in the environment units of Colombia and other countries, for example, say that are "discouraged" from acting against industries that dump waste into rivers. They instead seem to expend most of their energy on detaining small-time traffickers or abusive pet owners rather than acquiring the expertize, resources, and backing to crack down on serious environmental crime. Officials of Bolivia's Forest and Environmental Police (Pofoma: *Policía Forestal y de Medio Ambiente*) conduct regular raids on the open market of illegal forest and animal products in the city of El Alto, but can do little to dislodge the politically connected family networks that run the market.[7] States throughout the region do not just battle organized crime, in short, but abet it through their own cooperation, disorganization, and absence.

5. *Resources and Training:* From training to transport, resources for environmental enforcement are in short supply. Officers in environmental police units complain of a lack of materials, ranging from gloves to handle toxic chemicals and wild animals to chemical equipment to deal with ubiquitous water contamination. Ecuador's UPMA even lacks its own vehicles in the Amazon, and

its chief in Quito is "always asking" the Environmental Ministry (MAE: *Ministerio del Ambiente*) for help with transport and inspections. Similarly, Brazil has just 1400 federal environmental police to cover the vast Amazon along with the rest of the country, while many states' environmental units lack operational equipment, adequate intelligence, and data banks. In Colombia, "resources are scarce" as well, says the head of Bogotá's environmental police, which has just two vehicles and eight motorcycles.[8] In Bolivia, Pofoma has units in just four of the country's nine departments. Lacking vehicles, the 22 officers of the Louisiana-size La Paz department often borrow the single truck of the NGO *Animales S.O.S.* Bolivia's other national enforcement agency, the Forest and Land Authority (ABT: *Autoridad de Bosques y Tierra*) monitors two-thirds of Bolivia—and adminsiters patents and licenses—with a paltry annual budget of approximately $10 million and just 200 officials (a quarter of what is says it needs).

Such limits circumscribe enforcement's geographic coverage. A central rationale for separate environmental units is the strain of covering so much varied territory. But that potential cannot be met amid a perennial lack of technology, personnel, and funds. Brazil's Institute of Environment and Natural Resources (IBAMA: *Instituto Brasileiro do Meio Ambiente e dos Recursos Naturais Renováveis*) is in charge of environmental enforcement, backed by the military and police. But budget cuts stemming from Brazil's political and economic slump forced IBAMA to fire so many agents that their geographic coverage was reduced to pre-2008 levels, when there was only one employee per 3088 square miles in states such as Pará. The number of security officers working on the environment is also low, and the impact is perhaps greatest in the protected reserves that are a bulwark of the Amazon's protection. Yasuní, continental Ecuador's biggest national park and a World Biosphere Reserve of the United Nations Education, Scientific and Cultural Organization (UNESCO), has only 25 officials, who say they cannot properly monitor the area, even with NGO backing. In Bolivia, 26 federal protected areas covering 12.4 million acres have an annual budget of just a half million dollars and the number of guards patrolling them fell from 350 in 2008 to 317 in 2013.

6. *Legal Clarity:* Environmental enforcement, of course, requires viable laws to enforce. Despite the unprecedented growth of Latin America's environmental law, its essential attributes—from the clarity of its provisions to supportive regulation in its implementation—are often weak. A big source of confusion in environmental law is the network of law itself. In most countries, environmental regulations fan out through four overlapping areas of law: criminal, civil (non-criminal disputes); administrative (covering state agencies), and commercial (covering other natural resource use). Within each, regulations are further dispersed into many levels (from national to local) and categories, including agriculture; water; hunting and fishing; petroleum; waste; forestry; mining; health; transport; wildlife, organized crime; and of course the environment (usually in a General Environmental Law). Most illegal activities, as a result, are also dispersed. Deforestation, for example, may be defined as activities that "range from acts related to the establishment of rights to land, to corrupt activities to acquire forest concessions, and to unlawful activities at all stages of forest management and the forest goods production chain, from the planning stage, to harvesting and transport" (Tacconi 2008, 3). This innumerably unfolding list of abuses itself demonstrates the difficulty of environmental enforcement. As laws are formed and reformed, as a result, conflicts ensue. Much controversy surrounding the 2012 reform of Brazil's Forest Code came from its alleged intrusions on due process and property rights, for example, with similar debates arising amid the perennial Congressional revision of Ecuador's law (*Ley de Gestión Ambiental*) to address needs such as codifying vague "norms of environmental quality" (article 14). A simultaneous source of legal opacity is vague wording. For example, law 1333 of 1992, Bolivia's Main environmental law, has been overshadowed by 2010's Law of the Rights of Mother Earth (*Ley de Derechos de la Madre Tierra*). Like article 71 in Ecuador's constitution, it is a paradigmatic shift that gives nature legal protection and burnishes the country's image. Within Bolivia, though, the law is widely derided as pure "discourse" and a "setback" for having no specific legal obligations and a skeleton-crew of to implement it. And it has since been seen as negated by the more concrete Law 337 of 2013 (Law of Support of Food Production and Forest

Restoration, commonly known as The Pardon Law), which pardons illegal clearing in large estates between 1996 and 2011, with a requirement to restore just 10% of deforested areas (Article 5).

7. *Judicial Interpretation:* Within the judiciary, the breadth of environmental law often opens a Pandora's Box of approaches among officials. While advocacy has grown for applying criminal law in environmental cases, most judges apply administrative and civil law, which they say is easier for citizens to initiate. Like many analysts, they also believe that criminal law's higher level of proof—demonstrating a link between perpetrator and victim in court, such a tying a polluted body of water to a specific polluter—is considered nearly impossible. A further complication is the fact that state agencies are often responsible, complicit, or negligent. Can a company be liable for pollution if the state sewer system was faulty? Such questions are taken up in the field of green criminology, which pushes for environmental abuses to be treated as statutory criminal acts. Although much environmental crime arguably inflicts greater damage on more people and in more places for a longer time than other crimes, most of it is not legally recognized or prosecuted (see White 2008). Documenting specific harms through set judicial procedures, such as submission of physical evidence, is thus regarded as the only way to bring the full weight of the law against environmental abuse. This "harm principle" also help to see the larger "social, political and economic *context*" (Hillyard and Tombs 2007, 11) that generates such abuse, as detailed in this book.

Without such a vigorous approach, ongoing legal uncertainty blunts the potential force of the enforcement agencies described above. Even severe forms of pollution—in some countries, just 2% of the water is certified to be of good quality—that it is accepted rather than litigated. And civil law sanctions—fines, confiscations, permit suspensions, and obligatory repairs—are still used almost exclusively over penal actions. But even those measures are usually unenforced. Less than 1% of deforestation fines are paid in Brazil, according to reports, and officials in Bolivia estimate that over 5 million *bolivianos* in fines (about US$720,000) are delinquent. In several countries, forestry laws connect punishment to the amount of illegal timber seized. "We hardly see jail for a polluter," says Manaus's principal environmental prosecutor, since "the law is very soft [and] environmental abuse is not considered [criminally] punishable." In Ecuador, the 2000 penal code was the first one to include environmental crimes, but NGOs and UPMA assert that incarceration of "just one to three years is too little. Judges do not punish as they should." None of the multiple infractions country's mangrove areas, such as planting sugar cane, has led to anyone being imprisoned. Throughout the region, individual suspects just "pay the police," while companies also prefer to pay fines, since the balance is often in their favor. Jurisprudence is further limited by the lack of judges and courts to develop it. Bolivia's Agro-Environmental Courts (*Tribunales AgroAmbientales*) were established in the 2009 constitution (articles 186–198), for example, but remain just Agro: they are not yet empowered to address environmental cases. Peru has set up an environmental court in the Amazon, but its exhausted judge is also saddle with tax, customs, and market laws as well—with little time get the environmental training.[9] Such conditions illustrate environmental crime's true extent: perpetrators fade into the past while harms continue into the future.

8. *Policy:* It is difficult to enforce environmental laws when they clash with government policies. The broadest realm of such policies is macro-economic. As detailed in this book, most of the Amazon basin is open to exploratory blocks for hydrocarbon and mineral exploration, even in indigenous and protected areas. Nearly 83% of Peru's Amazon is covered by such blocks (RAISG 2012, 27). The catch-all justifications for such concessions are vague "national interest" clauses to approve exemptions to laws such as those for protected areas. And Latin America's biggest export industries—mining, agriculture, and fisheries—have the highest proportion of negative externalities. The region's mangroves, for example, bear the brunt of aqua-chemical contamination from a booming shrimping sector. Bolivian officials estimate that up to 70% of deforestation is related to soy plantations (Pabón 2013), not counting roads to transport Brazilian soy to the Pacific. In approving a road that would bisect the Isiboro-Sécure Indigenous Territory and National Park (TIPNIS), Bolivia's government ignored legal obligations to consult with its indigenous communities. The Environmental Vice Minister fired in 2011 for not approving the plan says that it demonstrates how much the environment is a "political instrument" in the country.[10] Brazil's

Transport Ministry encourages roads that the Environmental Ministry tries to block, while Colombia's Energy Ministry promotes mining that other Ministries try to stop. Ecuador's *Programa Socio Bosque* provides incentives for rural people to preserve the rainforest, but acquiring titles to new tracts from the Agriculture Ministry obliges settlers to cut down half of the trees (not to mention the approximately $300 that loggers will pay for large ones). In a withering critique of state policy, Venezuela's leading environmentalist wrote a chapter in this book that exposes the potential devastation of a gigantic mining project that would tear apart both the country's Amazon as well as its environmental legal system.

9. *Regulation:* Such policy eviscerates not just the law's broad protections, but the regulations that implement it. Poor regulation fuels land invasions, say many policymakers,[11] and makes compliance with the law "too difficult," add business owners. It also fails to control the burgeoning number of small-scale mining operations inundating the Amazon, as discussed throughout this book. In every country, the process of obstaining licenses is rife with corruption, cut corners, and exemptions. In Brazil, IBAMA officers in charge of licensing say that salaries, equipment, and personnel are always insufficient for the constant backlog. As Eve et al. note (2000, 254), it is "much cheaper and bureaucratically simpler to get permission to deforest an area than it is to get a forest management plan approved," and many sawmill operators declare that their logs come from a zone where logging is allowed, such as sites destined for hydroelectric dams. Uncertainty also pervades regulatory agencies responsible for them. Many Bolivian officials complain of a lack of standards to measure quantities and damages of industrial waste, particularly those in the 1997 Mining Law. As in nearly every country, as described in this book, rarely are Environmental Impact Studies completed, even for major projects. State fragmentation weakens standards further. For example, a bill to reform Bolivia's Mining Law would transfer environmental licensing to the Mining Ministry, which would, claim environmental officials, stick their agencies in the back seat.
10. *Society:* As with economic wealth, the myth of the Amazon as "a land without people for a people without land" dies hard. As described in many of the following chapters, migration to the

Amazon continues apace, encouraged by governments eager to relieve demographic and political pressures. It also helps reduce boiling tensions within highland Andean communities caused by the drought and melting glaciers of global climate change. But tensions are often just transferred to the Amazon, where new arrivals fight with settled peoples and everyone fights with each other over issues like land titles and oil contracts (Vásquez 2014). Bolivia's ombudsman estimates that 80% of societal conflicts—such as those over pollution from the giant Mallku Khota silver-indium mine—stem from natural resources.[12] Environmental enforcement is also weakened throughout the region by the lack of coherent and consistent public pressure. The impacts of national opinion polls and movements (Hockstetler and Keck 2007) quickly lose steam as they spread out into the Amazon's vastness. Many environmental NGOs are distrusted by citizens and specialists for being "co-opted" by the government, but at the same time rarely reap the benefits of such alleged cooptation. Bolivia has expelled many international environmental groups, for example, while an arrangement for NGOs to co-administer Ecuador's mangroves dissolved when it did not lead to the necessary logistical arrangements.

The Book's Plan: A New Vision

The Amazon's very name connotes a powerful force and a forbidding landscape. But it is, as this book shows, incredibly vulnerable. Even a small river-side gas light kills thousands of bugs that would feed other animals. Throughout the basin, that destruction is rolling out on a thunderous scale. But a well-planned combination of coordinated enforcement, political pressures, and policy innovation is also underway to gear up for the new battlefields in the war to save the region. This approach is creating new spaces in communities around the region, between the structures of the state and the free-wheeling frontiers of extractions, between short-term demands and long-term sustainability. Through their clear description and analysis of these efforts, the authors of this book develop the most robust, comprehensive, and updated framework for understanding the dangers to the world's biggest forest. They also push us to re-think the nature and potential of the state, the law, and politics on the local, national, and international stages. As they show, this coexistence of epic destruction and robust responses places the twenty-first century Amazon at the center of a full understanding of global environmental protection. If the Amazon is the world's lungs, this is how the world breathes.

Notes

1. The categories comprising this range include governmental responsibility for environmental protection; rights to environmental quality; procedural environmental rights; and individual responsibility to protect the environment (see Boyd 2013). All Latin American constitutions that refer to the environment for its *own* sake—rather than its human benefits—are in the Amazon.
2. Interview with Marlene Ayala Luna, Chief of Planning, La Agencia para el Desarrollo de las Macroregiones y Zonas Fronterizas (Ademaf), La Paz, 9 January 2014.
3. "Lawless rivers and forests," *The Economist*, 30 November 2013, p. 38.
4. *El Día*, 'Policía forestal recibe dinero de traficantes', www.eldia.com.bo, 7 February 2012.
5. Interpol and United Nations Environment Programme (UNEP). See UNEP report *The Rise of Environmental Crime*, June 2016.
6. Interview, Mario Aufiero, President, Associação dos Delegados de Polícia do Brasil-Amazonas, 6–9 December 2011.
7. Interviews with and Marco Antonio Morales Mendieta, Chief of Pofoma, La Paz and Merida Nogales, National Director, Pofoma, La Paz, 6 January 2014.
8. Interview, Sonia Santamaría, Subcomisario, Policía Nacional, Bogotá, 13 July 2011.
9. Author interview, Cesar Espiritu Peruto Carrero, Iquitos, Peru, January 19, 2017.
10. Author Interview, La Paz, 8 January 2014.
11. Interview with Carlos Viteri, President of the Biodiversity Commission, Congress of Ecuador, Quito, 23 January 2015.
12. Author Interview, Eugenio Mullucundo Cadena, Coordinator, Program of Indigenous People's Human Rights, Defensoría del Pueblo, La Paz, January 8, 2014.

References

Boyd, David R. 2013. *The Status of Constitutional Protection for the Environment in Other Nations*. Vancouver: David Suzuki Foundation.

Chomitz, Kenneth, and Timothy Thomas. 2001. Geographic Patterns of Land Use and Land Intensity in the Brazilian Amazon. Policy Research Working Paper 2687, World Bank, Washington, DC.

Division of Environmental Law and Conventions, 2012. Brazil's Environmental Crimes Law. New York: United Nations Environmental Program.

Eve, Evalice, Francisco Arguelles, and Philip Fearnside. 2000. How Well Does Brazil's Environmental Law Work in Practice? *Environmental Management* 26 (3): 251–267.

Hillyard, Paddy, and Steve Tombs. 2007. From 'crime' to social harm? *Crime, Law and Social Change* 48: 9–25.

Hochstetler, Kathryn, and Margaret E. Keck. 2007. *Greening Brazil: Environmental Activism in State and Society*. Durham: Duke University Press.

O'Donnell, Guillermo. 1993. On the State, Democratization and Some Conceptual Problems. Working Paper 192, Notre Dame, Kellogg Institute.

RAISG: Amazonian Network of Georeferenced Socio-Environmental Information. 2012. *Amazonia Under Pressure*. Bogota: RAISG.

Tacconi, Luca, ed., 2008. *Illegal Logging: Law Enforcement, Livelihoods and the Timber Trade*. London: Earthscan.

Vásquez, Patricia. 2014. *Oil Sparks in the Amazon*. Athens, GA: University of Georgia Press.

White, Rob. 2008. *Crimes Against Nature*. Portland, OR: Willan Publishing.

CHAPTER 2

Amazonia, Organized Crime and Illegal Deforestation: Best Practices for the Protection of the Brazilian Amazon

Franco Perazzoni

Abstract This chapter examines the vigorous and focused enforcement the Federal Police of Brazil against the wide range of mass environmental crime in the country's Amazon region, which covers 60% of the entire rainforest Basin (the other 40% is divided among eight other countries). It concentrates on organized crime, of which environmental violations are a rapidly growing proportion, with a focus on criminal networks of deforestation and mining in Amazonas State, the world's ninth largest jurisdiction whose protection of the rain forest is almost singlehandedly responsible for Brazil's GHG reduction and a bulwark against the region's "arc of deforestation."

Keywords Amazon · Brazil · Crime · Environment · Satellite monitoring

---

Translated by Mark Ungar.

F. Perazzoni (✉)
Brazilian Federal Police, Brasília, Brazil

M. Ungar (ed.), *The 21st Century Fight for the Amazon*,
https://doi.org/10.1007/978-3-319-56552-1_2

Introduction

The Brazilian Amazon is a region of continental proportions. With 4.2 million square kilometers, it covers approximately 49% of Brazilian territory (8.5 million square kilometers). It is larger than nearly all of Europe, as represented in Fig. 2.1.

This huge size has long made Brazil's Amazon a focus of the international community and nearly every debate on global conservation. The centrality of the region is due not just to the magnitude of natural resources it holds, but unfortunately also to the extensive environmental impacts afflicted by illegal timber cutting, particularly in the protected areas such as Indigenous Lands and Conservation Units. Deforestation of the Brazilian Amazon has been tracked closely since 1998 by the National Institute of Spacial Research (INPE: *Instituto Nacional de Pesquisas Espaciais*). Between August 2007 and July 2008, the institute found a total of 12,911 square kilometers of deforestation in the region. But after adoption of a series of coordinated actions among the Ministries of the Environment, Justice, Defense and the Staff of the Presidency of the Republic,[1] rates of observed deforestation since 2010 have been the lowest registered by INPE since the systems of monitoring began operating in 1988 (Fig. 2.2).

Despite this progress, the Brazilian Amazon has lost around 25% of its total original cover—a scenario even bleaker in light of the fact that the original cover was around 95% in the early 1980s. The rates of 2014

Fig. 2.1 Amazonia legal and Europe comparison. *Source* INPE (2016)

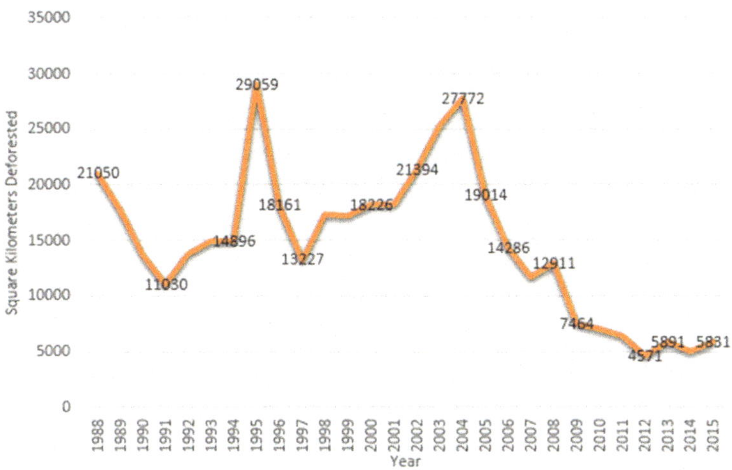

Fig. 2.2 Annual deforestation rates, Brazilian Amazon. *Source* INPE (2016)

and 2015, even after considerable reduction, still meant approximately 6000 square kilometers of deforestation—the equivalent of 600,000 soccer fields (90 m × 120 m). Among the resulting impacts are a loss of soil productivity; change in the hydrologic system[2]; the loss of biodiversity; global warming; acculturation of indigenous and traditional communities; tax evasion[3]; and rural violence.[4] Such data are alarming, particularly because of its implications for the world's climate and life. The situation appears even more intractable because illegal deforestation, the primary engine of the Amazon's destruction, is fomented in large part by powerful organized criminal groups, the focus of this chapter.

Illegal Deforestation as a Branch of Organized Crime

In 2010, after two years of investigations, the Brazilian Federal Police arrested some of the most important politicians and authorities for illegal logging in the Amazon. Operation Jurupari uncovered a massive illegal logging scheme that resulted in the arrests of over 90 people and caused an estimated $500 million in damage to the Amazon rain forest. Over 496 square kilometers (192 square miles), an area the size of California or

Spain, was affected, and an estimated 1.5 million cubic meters (642 million board feet) of timber was illegally extracted. While the investigation focused on activities in the state of Mato Grosso, which is located in the southern Amazon and is one of the most deforested states in Brazil because of clearing land for expanding soya farms, it also involved many others, including São Paulo, Paraná, Rio Grande do Sul, Espírito Santo, Minas Gerais, and the Federal District (Brasilia). The list of those arrested was a "who's who" of the forestry industry and included high-level government officials, business owners, and private landowners. Top officials in both the current and former Mato Grosso administrations were implicated, as well as those "throughout the State Environment Agency (SEMA), which oversees the logging industry" (Mongabay 2010).

When we speak of "organized crime" in Brazil, the first image that comes to mind is drug trafficking in cities like Río of Janeiro. But Jurupari clearly demonstrates that organized crime has definitively extended into the terrain of environmental crime. In fact, operations carried out by a range of state security forces, especially the Federal Police, have exposed a growing network of organized groups carrying out an expanding range of criminal activities against nature.[5] Empowered by their infiltration of state agencies, these groups focus on the extraction and illegal trade of forest products, minerals, and precious stones from the Brazilian Amazon. Brazil is the source of 75–80% of all timber exported from the region. Of that exported Brazilian timber, up to 70% is illegal. Such vast criminality is fueled by stratospheric prices of such goods and facilitated by the increasingly sophisticated strategies to avoid state control,[6] such as (a) contracting with people at different levels of the government; (b) the tight control, secrecy, and discipline in the communications and conduct of criminal operations; and (c) use of "persuasion" and recruitment, from corruption of public officials to threats against anyone who gets in their way—with acquiescence by public authorities, technocratic specialists, and professionals. In the area of illegal logging, common strategies are fraud in the authorizations of forestry concessions, forest management plans, and the document authorizing transport and commercialization of forestry products (DOF—*Documento de Origem Florestal*); the availability of fraudulent forestry credits that permit deforestation and illegal cutting in non-authorized zones, particularly in public lands and officially protected areas; and the transport, transformation, and commercialization of illegally source forestry products by sawmills and lumber businesses, which receive "legalized" products with fake documentation. These multiple forms of fraud are the fuel of a now vast criminal enterprise.

As discussed throughout this book, quickly catching up to logging as a source of destruction in the Amazon is legal and illegal mining. Like logging, illegal mining also involves a range of crimes[7]: (a) Fraud in the concession and acquisition of large-scale mining permits, or the extraction of minerals or gems with no authorized title at all; (b) Utilization of authorizations for a certain area with the sole intention to "legalize" illegally extracted minerals in other areas, especially public or indigenous lands and protected areas; (c) Falsification or alteration of documents, particularly adulteration of numbers on a permit to indicate a high permitted quantity, and so legalize extraction in regions where miners do have authorization by the National Department of Mineral Production (DNPM: *Departamento Nacional de Produção Mineral*)[8]; (d) Corruption of DNPM personnel to grant authorization, such as of Kimberley Process certificates (required for the transport and international trade of diamonds); and (e) Transport, commercialization, and export of those minerals, again accompanied by fraudulent documentation. Such conduct constitutes crimes already specified in Brazilian penal law, ranging from illegal tree removal, whose penalties range from one to three years of prison (article 38 of Law 9605-98), to more serious crimes such as corruption, falsification of public documents, money laundering and conspiracy, whose penalties together may exceed 30 years of incarceration.[9] Significant sectors from at least five major professions have lent their expertize and power to support, legitimize, and benefit from this criminality:

1. Engineers, topographers, and other technical consultants who apply their knowledge to approve state-funded activities and the completion of authorizations for timber forestry and mining operations.
2. Corrupt public officials, who are responsible for the approval of environmental projects, licenses, and authorizations.
3. Prospectors, power saw operators, and motor vehicle operators, who comprise the criminal pyramid's base. These individuals are generally recruited from local communities, receive the lowest

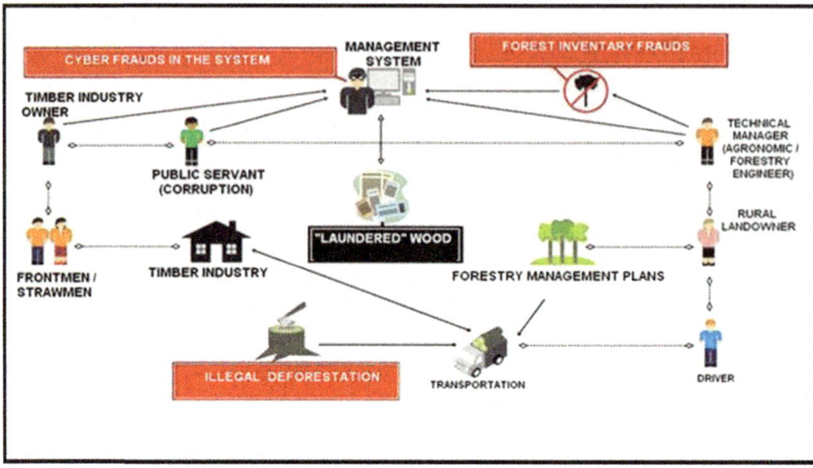

Fig. 2.3 Visual outline of illegal forestry in Brazil. *Source* Perazzoni (2012)

compensation, and, as the most visible face of illegal organizations, face the highest risk of detention.

4. Rural landowners, who are presumably responsible for any legal extraction of timber or minerals from within their property. Throughout the region, though, these landowners employ a very wide range of fraudulent arrangements to permit the legalization of minerals and forestry products extracted in illegal ways, not only from their own land, but also from within protected areas such as indigenous lands and national parks.
5. Wood and mineral companies, which have become the main economic pillar of this wide criminal enterprise. From protected areas throughout the country, illegally extracted timber and minerals are transported to these companies' facilities, from where they are later resold, duly legalized (Figs. 2.3, 2.4).

The first two of these groups, technical professionals and public officials, promote and provide key services to criminal enterprises. By issuing fake permits that legalize illegal activities, above all, they pave the way for criminals to expand the underground trade of forestry products and minerals. It is, therefore, no surprise that the World Bank (Pereira et al. 2011) urges Basin governments to better target the illegal extraction and

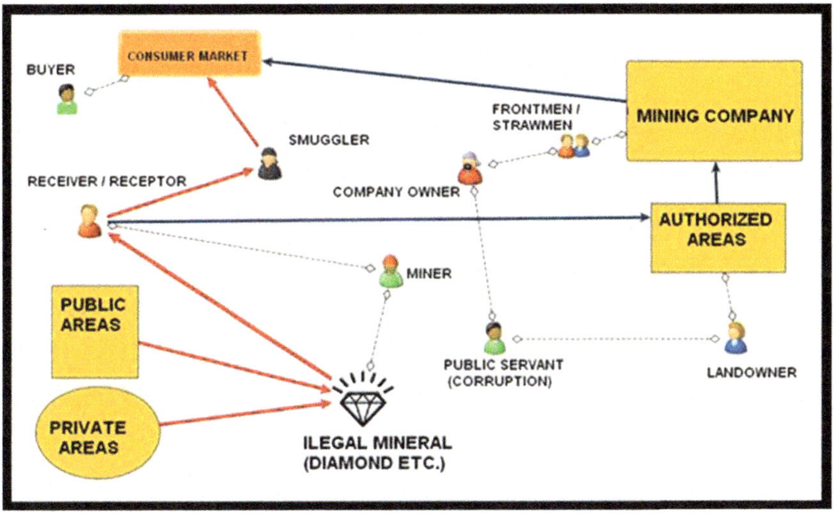

Fig. 2.4 Visualization of illegal mineral extraction in Brazil. *Source* Perazzoni (2013, p. 57)

trade of forestry products through better understanding of these spreading forms of white-collar collaboration. In particular, the World Bank pointed out that penal sanctions for deforestation in the Amazon are applied only to the weakest links—such as chainsaw operators, pickaxe workers, truckers, and miners—in the long illegal chain of criminality, leaving untouched the higher level and more lethal criminals and officials to whom most of the profit flows. It is, therefore, critical that state authorities start to adjust their strategies to focus on the nodes of these criminal networks. In this regard, Brazilian officials have made significant headway in the past several years, and that much of their approach can and should be applied to other countries facing similar situations. With a focus on this approach, the next section describes innovations and best practices—along with the lessons and ideas they generate—that are increasingly helping to fight illegal deforestation in the Brazilian Amazon.

Best Practices in Brazil in the Fight Against Illegal Deforestation

Sustainable Forest Management in the Brazilian Amazon

In 1965, Brazil adopted a Forest Code (Law 4771) that explicitly prohibits "the predatory exploration of the primitive forests of the Amazon basin" (article 15). As the code clearly states, the "exploration of forestry resources in that region can only be done in compliance with technical plans of conduction and forestry management, following the rules e comtent to be established by act of Government, within one year." Article 15 also requires logging activities to be "in accordance with technical plans and sustainable forest management, accomplishing with the regulations to be issued by the public authorities within one year." This law, in short, mandated a sustainable forest management for the country. But it was not until 2009 when the National Environmental Council (CONAMA: *Conselho Nacional do Meio Ambiente*) actually issued the first regulation (resolution n. 406-2009 CONAMA). That is, it took 44 years—not the one year required—for just an initial rule to be so established.

Prior to this regulation, Sustainable Forest Management Plans (PMFS is the Portuguese acronym) were prepared and carried out without following rules on the form and content of their forestry inventory—a document that identifies, locates, and quantifies all commercial species in the regulated zone, as well as establishment of the quantity of wood that can be extracted sustainably (since specified by CONAMA as 30 m^3/ha). In other words, any harvesting of native forests requires a PMFS, which stipulates the techniques of removal, forest replacement and management that are compatible with the area's ecosystems. A PMFS must comply with the following standards: (1) document the physical and biological environments; (2) determine the existing inventory of trees; (3) ensure that harvesting intensity is compatible with the capacity of the forest environment; (4) ensure that the cutting cycle is compatible with the time for reestablishing the volume of timber harvested from the forest; (5) promote natural regeneration of the forest; (6) adopt appropriate silviculture systems; (7) adopt appropriate forest harvesting systems; (8) monitor development of the remaining forest; (9) adopt mitigating measures for environmental and social impacts. An approved PMFS gives the holder authorization to practice sustainable forest management, but with submission of an annual report with information covering the management area and a description of activities conducted. This approach was created to enable continuous activity in a sustainable manner by dividing each area into smaller annual production units (UPAs—Unidades de Produção Anual). The number of UPAs in a Management Plan correspond to the cutting cycle, which is the number of years the harvested area needs for complete regeneration and be ready for harvesting again. When the last UPA is harvested, harvesting can return to the first one the following year.

Since the circulation of wood in Brazil is controlled by the government through computerized systems (SisDOF, discussed below), the approval of a PMFS immediately generates forest credits that correspond to the species and volumes of timber authorized for harvesting. These credits are placed in the account of the respective holder and, as the timber is harvested and sold, a DOF[10] is issued. Accompanying the timber all the way to the buyer (usually a lumber mill), the forest credits corresponding to the timber shipped will be transferred from the seller's account to the buyer's account. Although Brazil's forest management structure remains rife with the irregularities and the types of fraud mentioned above, significant and durable advances have been made under a new model. Among the main reasons for improvement is that data from all these businesses is centralized in state agency databases, allowing officials to conduct audits and analysis, even remotely, on the activities carried out under any PMFS, particularly regarding preventative measures and appropriate restrictions. Table 2.1 below spells out the different stages of development and approval of a PMFS.

We now turn to a focus on controls of timber and mining.

Systems to Control of Timber and Mining Activities: SisDOF, Cadastro Mineiro, and SIGMINE

Up until 2006, an Authorization for Transport of Forest Products (ATPF: *Autorizacão para o Transporte de Produtos Florestais*) a document printed on specialized paper by the Brazilian Environmental and Natural Resource Institute (IBAMA: *Instituto Brasileño del Medio Ambiente and Recursos Naturales Renovables*), the country's top federal environmental agency, was required for any business carrying out extraction, storage, commercialization and transport of wood and other forestry products. This system was practically entirely manual, which not surprisingly opened a very wide door to fraud. A small cottage industry even sprung

Table 2.1 Phases of development of a PMFS in the Brazilian Amazon

Stages	*Activities*
Zonification 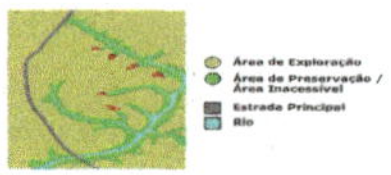 	During zonification, the rivers, lakes, and roads in the property are identified. During this stage, the zonification divides the forest property in explorable zones; permanent preservation areas (APPs); and non-exploitable zones (riverside areas and other courses of water and vegetation as steep slopes or hills)
Forestry inventory 	Documentation of vegetation and general conditions of the forest through a forest inventory. Species that are commercialized are inventoried, as are the types of trees that, by legal imposition, may not be felled (particularly protected species and seedlings, which are used for natural regeneration. Since seedlings serve as the gene source for the new crop, they must be phenotypically superior, prolific in flowering, sturdy, and disease-free)
Road planning 	Main roads are planned, as well as secondary roads that will connect the exploration area with principal roads
UPA división  	The assigned areas are divided into smaller UPA units, each of which corresponds to an annual cutting cycle that must be followed. The number of UPAs in a Management Plan should correspond to the cutting cycle, which is the number of years the harvested area needs to completely recompose itself, and be ready to be harvested again, in such a way that, when the last UPA is harvested, it is possible to return to the first one the following year. The most common cutting cycles are 10, 20, or 30 years
Authorization 	With permission of the environmental agency, the PMFS account is created in the Electronic Forest Management System (SisDOF). This account is analogous to a bank account: forestry credits are equivalent to the volumetric and species of the forestry inventory, that will be deducted with each sale, through the registry of the operation in question in the system and the emission of a transport authorization that must accompany the product to its destination (DOF)

(continued)

Table 2.1 (continued)

Stages	*Activities*
Tree harvesting 	Once the PMFS is authorized by the environmental agency, designed trees may now be felled and sold, but in clear accordance with the established demarcation of annual units (UPAs) and the forestry inventory data. Harm to protected species is strictly prohibited to harvest seed trees and those with a diameter inferior to the environmental regulations in force

Source Perazzoni (2013)

up of groups formed exclusively for the sale and transport of forestry products utilizing “ideologically” false ATPFs (a valid document with false information ATPFs stolen from IBAMA offices, or counterfeit documents made from scratch.

Because of these weaknesses, ATPFs were replaced in 2006 by the Electronic Forest Management System (SisDOF), a non-paper control of sales, purchases, transport, and storage of forestry products and sub-products. A document authorizing such activity (DOF) is done through IBAMA’s website by prior registration of the forestry specialist or company, which must itself be registered and legally authorized, ensuring there are no pending questions over their status or activities. Upon approval of its PMF, a “forestry credit” account is put into IBAMA’s SisDOF database. These credits, in turn, are adjusted in the SisDOF account with each transaction of the respective buyer (the company) through a transit letter that must accompany the product and be presented to officials at any inspection. The system thus forces the buyer to demonstrate the legal origin of all products purchased and in their stock. It also documents changes in ownership of a product and any transformation of the product itself (such a cutting timber into boards) as it moves along the supply chain. Through that approach, the DOF tracks every native-origin forestry and sub-forestry product until its destination by any form of transport: road, air, railroad, river, and sea (Fig. 2.5).

In short, the DOF is a required license for movement of all native-origin forestry and sub-forestry products, including native vegetal charcoal. SisDOF electronically documents and controls the balance (credits

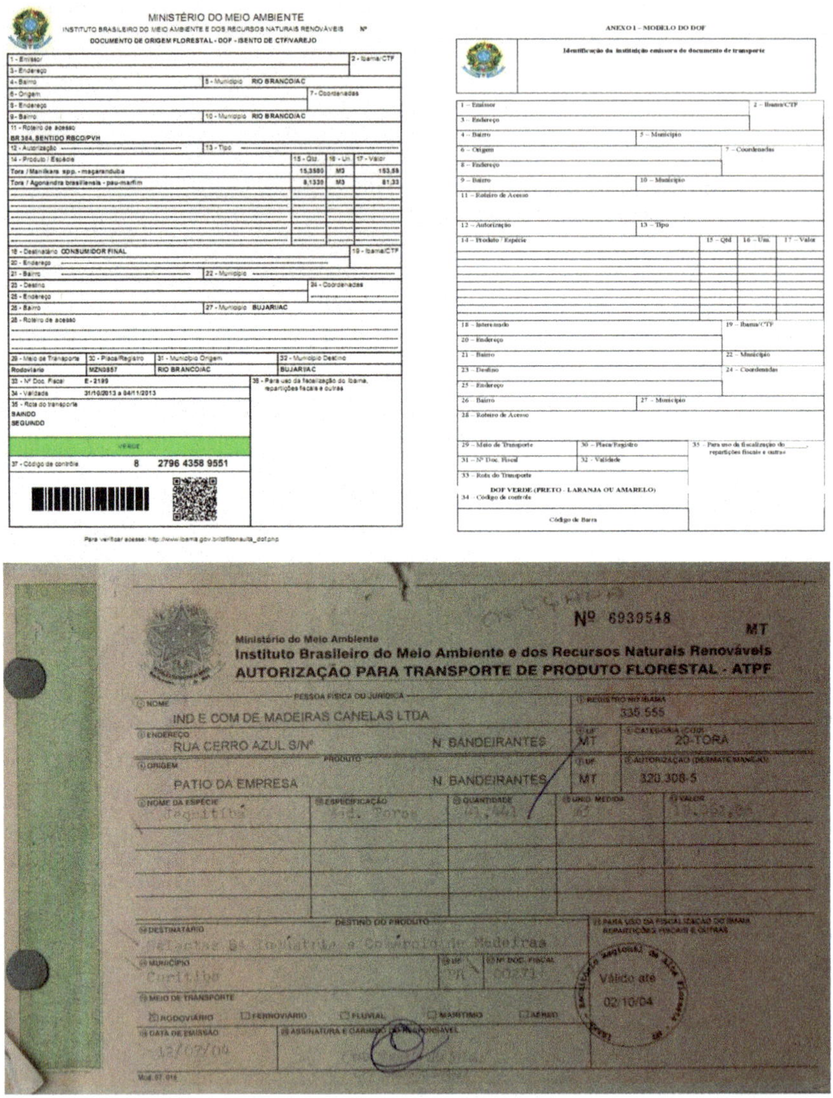

Fig. 2.5 Examples of an ATPF and DOF documents

and debits of the forestry products in the accounts of each participant in the supply and trade chain operations) in the same way as a bank tracks its customers' accounts. The issuance of a DOF is like a cheque from a bank account: it requires credit for the wood or forestry sub-product in relation to the emitter; these credits are then transferred to buyer's SisDOF account. Such careful tracking has provided much stronger control of the entire chain of ownership of forestry products, from extraction to sales or export. In contrast to the ATPFs, SisDOF makes this possible through its comprehensive database, which has for all commercial transaction the type of product, species, volume, origin, destination, and transit, as well as data on the transporting vehicle, the time in route to the recipient, and the sales receipts, among other documentation. This thorough approach, though, has not made SisDOF completely immune to the well-honed ruses of unscrupulous operators and of groups specializing in environmental crimes that operate in Amazonia's many corners. Because criminality extends beyond SisDOF's scope, its needs to be supported by tele-detection and geo-intelligence data, particularly from satellite imaging in the zones of operations, to carry out analysis and data mining to more fully and accurately detect and identify the wide range of fraudulent schemes or irregularities that previously went unnoticed by the authorities.

The two government systems established to provide that support are *Cadastro Mineiro* and SIGMINE, which are accessible through the Internet and maintained by the National Department of Mineral Production (DNPM: *Departamento Nacional de Produção Mineral* at www.dnpm.gov.br). *Cadastro Mineiro* is a conventional database that permits access to registry data and authorization to any business or person authorized to carry out mining activities anywhere in Brazil. The specific data includes the location of the authorized mining area, minerals allowed to be extracted, the technician overseeing the operation, the specific types of activity, the valid dates of each mining and environmental permits. Within this system, searches can be conducted by a wide range of criteria, including the number of the DNPM case, the name of the company or its owner, the municipality, or the substance being extracted. Such searches then verify the accuracy of the licenses and if the specific person or business is in fact authorized by state agencies to carry out mining activities. The complementary SIGMINE is a Geographic Information System[11] (GIS), also accessible through the Internet that allows, based on case number or the coordinates of a

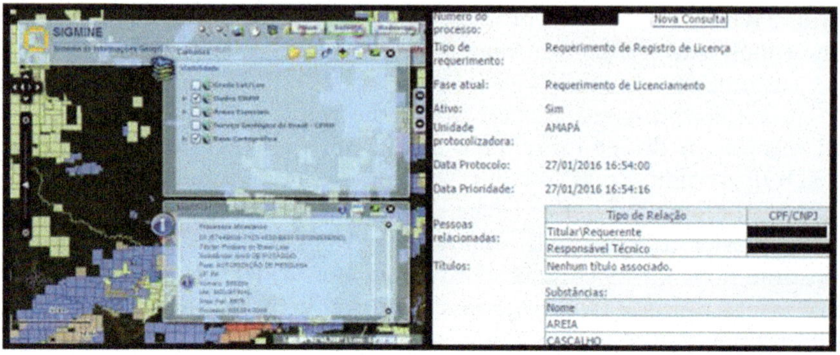

Fig. 2.6 Screenshots of a SIGMINE and Cadastro Mineiro search. *Source* DNPM (2016)

specific geographic location, to rapidly and securely identify the location and dimensions of the pertinent estate in the authorized zone.

The joint use of *Cadastro Mineiro* and SIGMINE has been extremely useful to officials by allowing them to not only verify if a particular person or company has the permits for mining activities, but also to determine, based on SIGMINE data (which can be downloaded onto any device or cell phone with GPS), if the physical limits of the authorized area are being respected by the owner of the extraction permits (Fig. 2.6).

Systems of Official Monitoring of Burning and Deforestation

Brazil uses three powerful systems to monitor deforestation in real time: DETER (Real-Time Detection of Deforestation), PRODES, and PROARCO. DETER is an alert system developed by the National Institute of Spacial Research (INPE: *Instituto Nacional de Pesquisas Espaciais*) to support both domestic and international organizations. Its data is geared toward the formation and prioritization of both preventive and policing actions against illegal deforestation, such as with financial penalties, administrative fines, and criminal investigations. Data is gathered and reported on a monthly basis for deforested estates of over 25 square hectares. Through its real-time gathering of this information, accessible at www.obt.inpe.br/deter/, DETER facilitates a wide range of research and analysis: searches for deforested areas, for example,

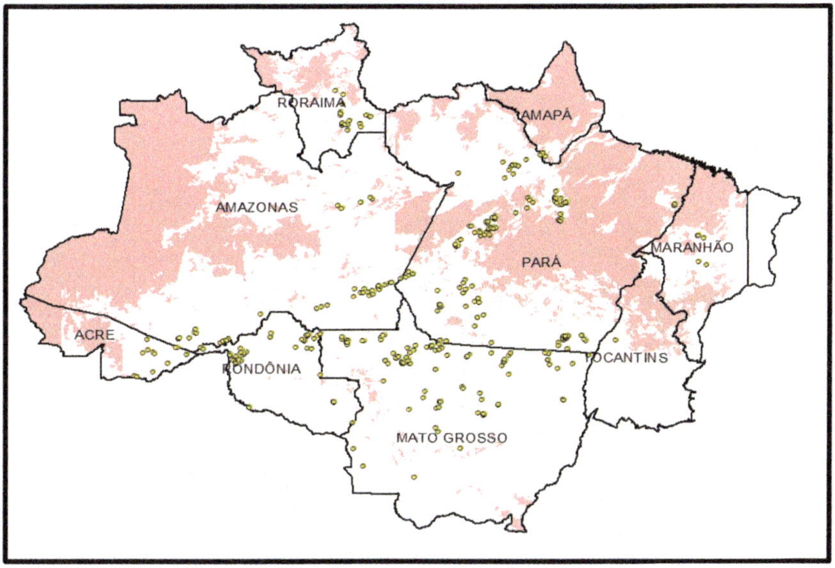

Fig. 2.7 Map produced thorough data of DETER in January 2016. The *yellow dots* indicate the location of deforested areas; the *red areas* represent cloud cover. *Source* INPE (2016)

can be conducted specifically within the interior of either federal and state-designated conservation areas; within protected indigenous lands; or by state, city, or more generally, for the entire Brazilian Amazon (Fig. 2.7).

Since 1988, Brazil has used a system called PRODES to measure annual rates of deforestation (a PRODES year begins in August and ends the following July) to detect deforestation of areas over a 6.25 hectares in the entire Amazon region (If someone deforests one ha or less, it will not appear on PRODES—a problem afflicting deforestation enforcement throughout the Basin). The initial presentation of data is given in December each year, as estimates, while consolidated data is presented in the first semester of the following year. Access to and format of the data are similar those of DETER and are done though the Internet page at http://www.dpi.inpe.br/prodesdigital/prodes.php (Fig. 2.8).

For over 20 years, INPE has also maintained an operative satellite system of forest fires, known by its acronym, PROARCO: www.dpi.inpe.

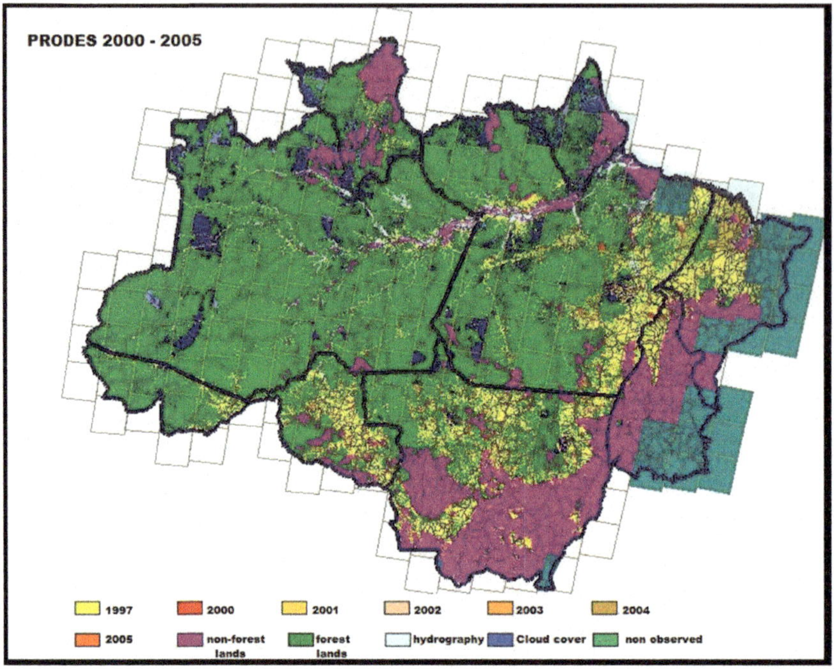

Fig. 2.8 Cumulative map, PRODES 2000–2005. *Source* INPE (2016)

br/proarco/bdqueimadas. The data that PROARCO produces and publicizes is not limited to Brazil or even South America—it also provides up-to-date information on forest fires on different continents, including Africa and North America. Such data can be obtained through a search by region or by direct navigation on the map on the project site (Fig. 2.9).

Use of Geo-Intelligence and the GAGeo Project of the Federal Police

In addition to well-known means of criminal investigation such as communications tapping (e.g., telephone and e-mail) and banking or financial transactions (e.g., checking accounts and credit cards), the growing skill and efficient use of geo-intelligence (GEOINT) techniques have

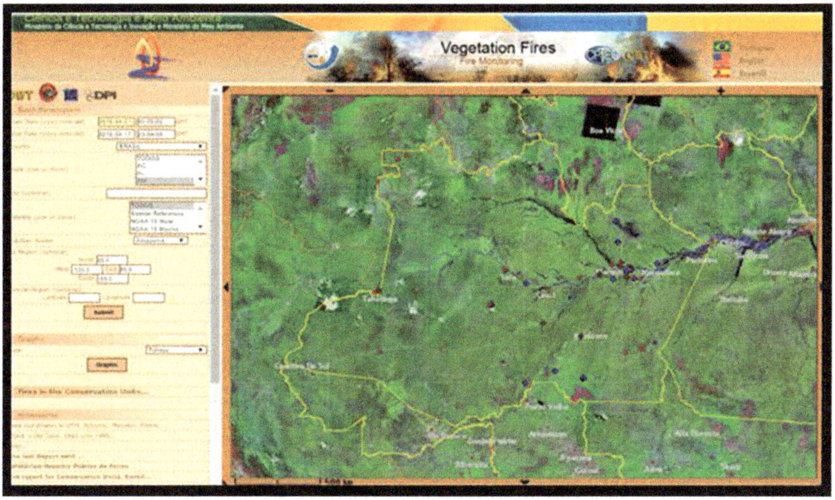

Fig. 2.9 A PROARCO search and map (*Fuente* INPE 2016)

become indispensable to Brazilian penal authorities, who struggle with the huge physical dimensions and complex characteristics of the Brazilian Amazon. The use of images obtained by satellites and airborne sensors (from a plane, helicopter, or, increasingly, a drone), along with the evolving capacities to analyze spatial and non-spatial information, has led to breakthroughs in the detection of fraud by criminals who would otherwise be impossible to detect, even with extensive human and material resources. To deepen this work, the Federal Police's Environmental Crimes Division (*Divisão de Repressão a Crimes Contra o Meio Ambiente*) established its Analysis and Geointelligence Group (GAGeo: *Grupo de Análise e Geointeligência*) in 2010. Since then, GAGeo has launched a range of operations and police investigations in the Amazon, many of them utilizing publicly available software and data. Its success with such instruments has led to global recognition, such as by the Association of Police Forces of the Americas (AMERIPOL 2014).

To best appreciate GEOINT's impact, it is first necessary to understand how it works. The most common definition of GEOINT is the one developed by the National Geospatial-Intelligence Agency (NGA) and used in US legislation: "[...] the exploitation and analysis

Table 2.2 Components of GEOINT

Imaging	Visual registry of objects (both naturals and artificial), produced by satellites, aerial platforms, unmanned aerial vehicles (UAVs), or through related means
Intelligence of images	Technology to extract useful information through interpretation or analysis of imagery and collateral data[a]
Geo-spacial information	Information on the surface of the Earth that identifies the location, geometry and attributes of the surrounding, buildings, objects, resources, or phenomena that occupies physical space (which may be base on maps, statistical data, graphs, archives, and digital data)

(NGA 2006)

[a]Collateral data is all non-spatial data of different types that may contribute to the understanding and interpretation of images, including the information of data and intelligence from other sources

of imagery and geospatial information to describe, assess, and visually depict physical features and geographically referenced activities on the Earth […] GEOINT consists of imagery, imagery intelligence, and geospatial information." (Title 10 U.S. Code §467). More specifically, GEOINT comprises the following components (Table 2.2):

Such categorization is of course conceptual, since practice necessitates the combination of them that more precisely captures GEOINT's principal function: consolidating all activities involved in the planning, acquisition, processing, analysis, and diffusion of spatial data to allow for the fullest possible comprehension and contextualization of information from all intelligence sources (Fig. 2.10).

GEOINT was initially conducted primarily through aerial photographic recognition and interpretation of the resulting data was completely manual. But it developed rapidly. Currently available is a wide range of resources and systems of targeted information, most commonly and broadly called Geographic Information Systems (GIS), which enable us to work with large quantities of digital data from many sources, including satellite images. This chapter now turns to the ways in which GEOINT can help identify and prosecute illegal deforestation crime in Amazonia. It will center on a case that illustrates GEOINT's foundation and potential by showing how it used satellite images to confirm extensive of activities of mining between 2010 and 2011 in a specified area (Fig. 2.11).

Although these dual photographs clearly show intensive mining activity in the areas in 2010 and 2011, they do not show if the extraction

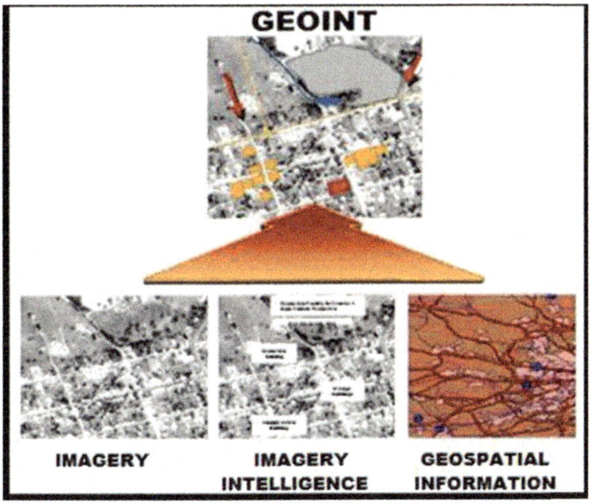

Fig. 2.10 GEOINT components NGA (2006)

was legal or identify those conducting it. Such photographs must therefore be analyzed with additional information from other sources, such as environmental agencies, DNPM and the Federal Police of Brazil. In this case, using the geographic coordinates of this mining operation, a search was conducted with the above-mentioned SIGMINE and Catastro Minero systems. Examining the region, the search identified the existence of a process in the DNPM and obtained the digital archive of a specific polygonal of territory. With that material, it was able to discern that there was a business located within it, but that the business did not have a valid permit for the exploration of minerals emitted by DNPM, since August 2007 (Fig. 2.12).[12]

In short, analysis of all this data (satellite images, the geology of the polygon, and the information recorded in its DNPM case) enables verification not only that the mining being conducted in the zone was clearly illegal, since it took place outside the prescribed limits, and at a time much later than August 2007, after when no document had authorized mining under any pretext or legal title (Fig. 2.13).

Obviously, a complete clarification of this case would require more detailed analysis, as well as the realization, in situ, of investigation activities such as technical examinations lead by forensic experts to identify and evaluate the environmental damage. But this example still provides

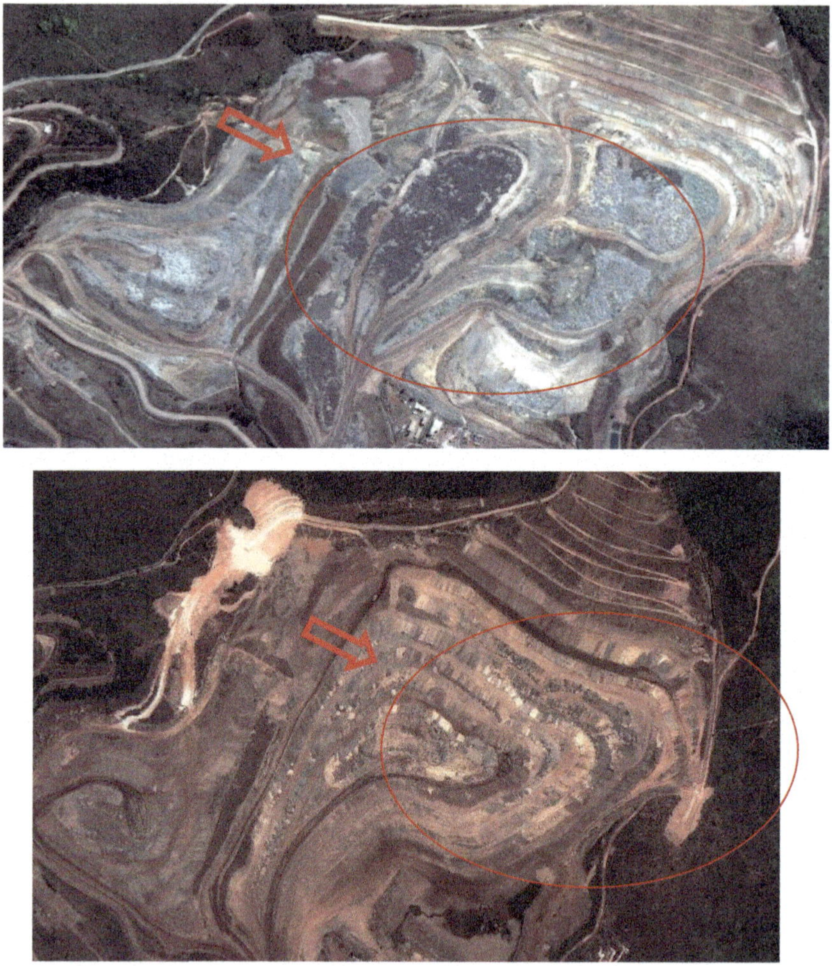

Fig. 2.11 *Above* Orbital image in 2010 of a mining zone. *Below* Image of the same zone in 2011 (GAGeo 2016)

a general vision of the many opportunities of GEOINT for the investigation of crimes committed against the environment. After all, without the joint analysis of the spacial and non-spacial data that we used in this system, any verification of the infraction would have been difficult,

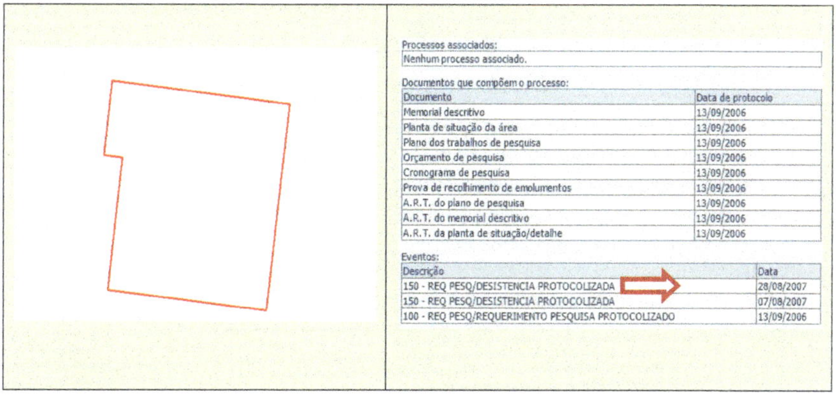

Fig. 2.12 Mining area polygon file and information available in the *Mineiro Cadastre* (GAGeo 2016)

Fig. 2.13 Joint analysis of vectoral data, the image, and the DNPM mining case (GAGeo 2016)

in particular in operations carried out illegally after 2007. Without evidence or *in flagrante delicto* seizures, perpetrators can always argue that the mining was done years earlier, prior to the expiration of the DNPM license. In other words, the joint processing and analysis of all this data allow us to confirm the practice of crime in this specific area and time (Imagery + Imagery Intelligence + Geospatial Information = GEOINT).

As mentioned above, in 2010, the Federal Police's Division of Crimes Against the Environment launched a project called GAGeo (Portuguese acronym for Analysis and Geointelligence Group), which brings together specialized agents to carry out analysis and GEOINT on behalf of the Federal Police for investigation of groups conducting illegal deforestation, especially in the Amazon. In recent years, the activities of GAGeo provided the data, information, and analysis that enabled a range of police operations, many of them undercover.

Among the operations that have been made public, particularly illustrative was Operation Salmo 96:12, carried out on May 23, 2012. It began with an investigation by the Federal Police in Roraima State, which had gathered evidence of serious fraud against the federal land agency, the National Institute for Colonization and Agrarian Reform (INCRA: *Instituto Nacional de Colonização e Reforma Agrária*), in the unlawful acquisition of property in the State. Investigative activities, such as telephonic interception supported by judicial warrants, also exposed corrupt public officers involved. In the course of the investigation, suspicion arose not just of illegal acquisitions through fraudulent documents, but the spread of environmental crimes like unauthorized deforestation within both the zones under investigation as well as nearby protected areas. Initially, GAGeo agents thus carried out exhaustive analysis of the all trade of forestry products within and from these illegally acquired properties. We found that nearly all these properties' owners had licenses to carry out cutting in the forests and had sold large volumes of wood to several sawmills in the area in a short amount of time. But also found through SisDOF analysis were various inconsistencies indicative of fraud in the movement of forestry credits for the legalization of wood extracted in other areas. Around 40 authorizations given by the state environmental agency were identified, and all of them involved some kind of fraud of SisDOF during the process. In all these cases, the data analysis, along with satellite images of the region and vectoral data defining property boundaries, allowed not just confirmation of fraud on forestry credits, but other environmental crimes as well, such as unlawful extraction within areas of permanent preservation, and deforestation within properties of areas much larger than those authorized by the environmental agency.

To show how such crimes are conducted, the next section turns to another emblematic case: the Property Known as "Ranch 1." Covering 1237,108 hectares, this property was located in the municipality of Rorainópolis, approximately 12.2 kilometers from the Anauá National Forest and 50.8 kilometers from the PA Ecuador (PA is a

Fig. 2.14 Landsat-5/ TM Image October 15, 2010 shows no deforestation within the property perimeters (*blue polygon*)

Projeto de Assentamento, or Settlement Project, which is a federal public area in the Amazon destinated to land reform). Both of those areas are under federal jurisdiction. The preliminary analysis of this property's SisDOF account revealed several incompatibilities, especially regarding vehicles supposedly used for the transport of forestry products (such as passenger vehicles, utility vehicles, and motorcycles).[13] Interpretation of satellite images of the region along with SisDOF data, however, brought in new and valuable elements to the investigation:

The Landsat-5 / TM 15/10/2010 image below shows that, on October 15, 2010, there were no signs of deforestation within the perimeters of the property, or that could be confirmed by environmental agents of IBAMA who inspected that area on March 29, 2011 (Fig. 2.14).

However, SisDOF registered the sale of 4486.567 m^3 of wood from different forest species from January 20, 2011 to March 24, 2011—that is, in the period prior to this image and IBAMA's inspection. This showed that all of those DOFs were issued to cover up forestry products originating in other non-authorized zones. Following the issuance of the DOFs, as well as the IBAMA agents' inspection and of the respective administrative fine by federal environmental authorities, the Landsat-5 Imagen/TM of August 31, 2011 revealed approximately 489 ha of logging, which exceeds the 212 approved. It also verified that the exploration disregarded the limits established in the project as well as by environmental norms. In total, approximately 276.7734 hectares were illegally deforested (Fig. 2.15).

Because this property was fraudulently acquired through the state agency in charge of agrarian reform and that, therefore, all of the deforestation within it was illegal, the perpetrators were subject to penalties for document fraud and illegal deforestation. They were also required to provide economic compensation for the environmental damages. The forensic expert assigned to the case estimated that the cost to repair these areas combined with the total value of the illegally produced forestry products was R$ 9.546.714,46 (US$4 million at the time).

In the Salmo 92:12 case, criminal evidence was acquired through GEOINT techniques, which exposed a network of fraud and corruption that covered up deforestation in Roraima with an appearance of legality. Such fraud was carried out in two phases: (1) through submission of false data and documents to INCRA, acquisition of property titles for public areas designated for land reform in the state; and (2) acquisition of permits to cut down trees in these areas, along with fraud and the corruption of public officials in both state and federal environmental agencies.[14] Analyses conducted through GAGeo demonstrated that, in addition to carrying out illegal deforestation within areas where permission for activity was fraudulently acquired, the criminals also conducted false transactions with the corresponding forestry credits, with the aim of legalizing forestry products extracted in other zones.

GEOINT data and information, together with other penal mechanisms to obtain proof (such as monitoring, stakeouts, telephonic surveillance and information obtained through bankruptcy and financial records), laid the path for Operation Salmo 96:12, in which arrest warrants were carried out against 44 suspects, including eight officials in IBAMA, three in INCRA, and six of FEMARH, Roraima State's environmental agency. Data information and penal proof produced by GAGeo also allowed the police to attain, through the courts: (I) dismissal of 18 public officials; (II) confiscation and destruction of all the good (moveable and stationary) of 28 individuals and 20 businesses, in addition to suspension of all forestry permits that were fraudulently issued to them. In total, 115 persons were charged for crimes of land invasion and illegal deforestation. In addition to its scope, what was notable about this case is Roraima state historically has not been one of the major areas of Amazon deforestation. But just between 2011 and 2012, it experienced a 368% increase in deforestation, much of which

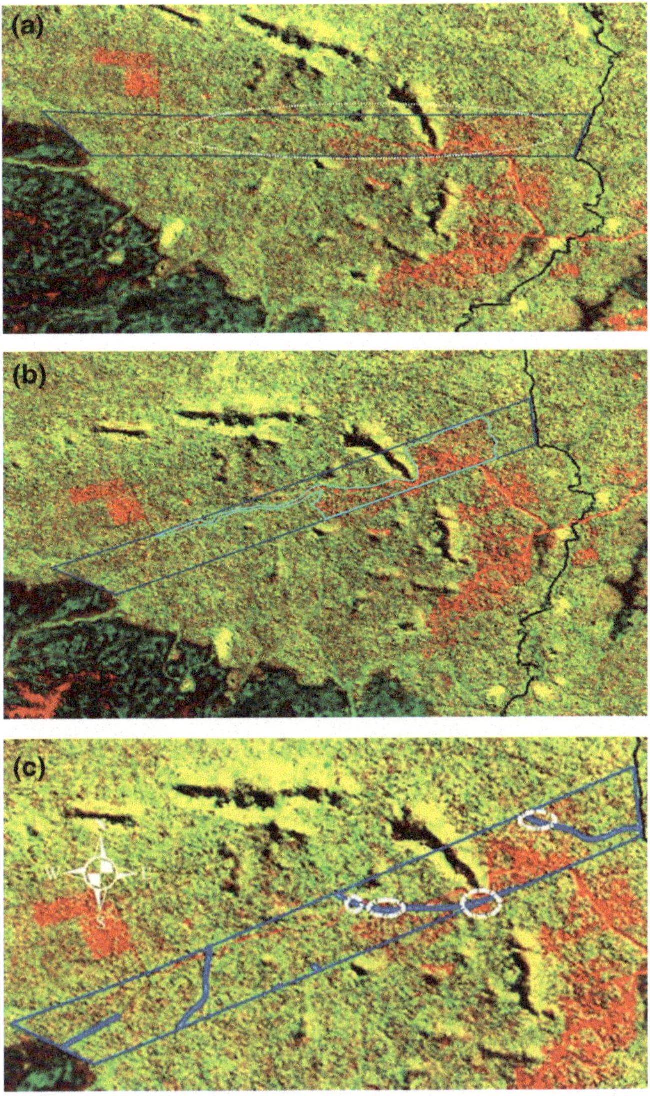

Fig. 2.15 **a** Landsat-5 Imagen August 31, 2011 shows the cutting in a significant part of the rural property (*reddish areas* within the *white ellipse*) and its outskirts in the date after issues of the DOFs. **b** The same image revealed that deforestation within the property was approximately 489 hectares. **c** The image also showed that deforestation within the property neared 15.53 hectares of Areas of Permanent Preservation (*APP* Áreas of Preservación Permanente)

was certainly connected to the group investigated by this operation. That realization only reinforced GEOINT's importance of as an instrument of protection in the Amazonia.

Special Operations Against Illegal Deforestation in the Amazonas

In the past 5 years, the Police Federal of Brazil has conducted a range of special operations against illegal deforestation in the Amazon.[15] The legal mechanisms that they use include Quebra de segredo, literally translated as "breach of secrecy," which is judicial authorization to access data or information that is otherwise confidential, like bank and financial accounts, as well as to intercept communications like phone and e-mail exchanges. Another primary mechanism is Entrega controlada, or "controlled delivery," used when a consignment of illicit products (drugs, timber, minerals) is detected and allowed to go forward under the surveillance of law enforcement officers in order to identify all those involved. Some of the main operations, as well as their results, are as follows (Table 2.3):

Table 2.3 Principal special operations against illegal forestry and mining in Amazonía, 2010–2015

Operation	*Date*	*Objective(s)*	*Search orders*	*Detentions*
Jurupari	May 21, 2010	To investigate fraud in forestry management plans and corruption of public officials related to the extraction, transport and illegal sales of forestry products from federal protected areas in Mato Grosso state, such as indigenous lands and national parks	94	91

(continued)

Table 2.3 (continued)

Operation	*Date*	*Objective(s)*	*Search orders*	*Detentions*
Ouro Preto	June 16, 2011	To dismantle a criminal organization whose illegal mineral extraction operations were devastating Amazonia's tropical forests. Investigation began in 2008 of illegal extraction of cassiterite (tin oxide) within the Indigenous Land Tenharim-Igarapé-Preto, located in a remote area with difficult access, straddling Rondonia, Amazonas, and Mato Grosso states	18	8
Salmo 96:12	May 23, 2012	To dismantle a sophisticated criminal apparatus responsible for the 363% increase in deforestation in Roraima state in 2011 and 2012. It identified participation of 24 businessmen, eight IBAMA officials (including Roraima's inspections chief); three in INCRA, six in FEMARH; several in the Roraima state Land agency (ITERAIMA), one municipal official, and over 100 front men	06	44

(continued)

Table 2.3 (continued)

Operation	*Date*	*Objective(s)*	*Search orders*	*Detentions*
Xawará	July 13, 2012	Against environmental crimes and illegal exploitation of gold in the indigenous land of the Yanomami. Confiscated were eight (08) kilograms of gold, six aircratft, and US$150,000	44	33
El Dorado	Nov 6, 2012	To dismantle a criminal network of illegal gold extraction in the interior of the indigenous land of the Kayabi and illegal mining in the Teles Pires river region of Mato Grosso state	64	28
COOPI	Dec 20, 2012	Operation to halt irregular tree cutting in indigenous lands in the municipalities of Maranhãozinho and Centro do Guilherme in the state of Maranhão	7	0
CONTRA GARIMPO	March 3, 2013	To combat the actividad minera ilegal en Pontes e Lacerda, MT, Results: Confisaction of three trucks; two chargers; one tractor; two motorcycles; one backhoe; one scale to weigh gold; 246 grams of gold, and 458 grams of mercury	01	05

(continued)

Table 2.3 (continued)

Operation	*Date*	*Objective(s)*	*Search orders*	*Detentions*
Curatetinga	March 22, 2013	To combat illegal gold mining in the indigenous land Ynomami, near the Uraricoera river in Roraima. Results: Confiscation of 8000 L of combustible diesel, 3000 L of gasoline and canoes to transport equipment and people; three arms, 10 chainsaws, motors, and other mining equipment	01	42
Dríade	April 3, 2013	Combat illicit business groups conducting illegal deforestation in the Biologic Reserve of Gurupi, in the far north of Maranhao state. Wood extracted illegally from the reserve was taken to zones where the group did have authorization for cutting and production. DOFs and other documents issued by environmental authorities were utilized to "legalize" wood that was then exported	29	0

(continued)

Table 2.3 (continued)

Operation	*Date*	*Objective(s)*	*Search orders*	*Detentions*
Ibira	May 21, 2013	To fight illegal extraction of wood in the indigenous land of Alto Río Guamá (with a surface area of 280.000 hectares, in the municipalities of Paragominas, Piriá, Norte Garrafão, and Santa Luzia do Pará;) and fraud in the SisDOF	27	0
Termitas II	Nov 6, 2013	To combat criminal organization that carried out fraud in SisDOF and the illegal extraction of wood in the interior of protected areas of Pará state.	21	07
Trairão	Nov 20, 2013	To combat environmental crimes and fraud in the use of forestry credit (SisDOF), the main objective of Operation Trairão was to halt illegal timber in the west of Pará state	03	01
Nuvem Negra	Dec 3, 2013	Taking place in the states of Maranhão, Pará, and Goiás, it was a part of a full-year investigation spurred by detection of fraud in IBAMA's forestry control system (SisDOF)	07	43

(continued)

Table 2.3 (continued)

Operation	*Date*	*Objective(s)*	*Search orders*	*Detentions*
Kalupsis	May 7, 2014	To combat illegal timber in Aripuana and Conselvan, in the northeast of Mato Grosso state	67	9
Castanheira	Aug 27, 2014	To dismantle criminal organizations specializing in land appropriations and environmental crimes	22	4
Mesclado	Dec 10, 2014	To combat illegal felling in indigenous lands in the Southern Cone region of Rondonia	20	7
Xilofogos	Dec 18, 2014	To combat the deforestation in the state of Roraima	49	0
Filão do Abacaxi	Sep 18, 2015	To combat criminal organizations responsible for illegal mining in the Filão do Abacaxis region of Amazonas state	4	3
MAE Do Ouro I E II	Oct 28 and Nov 24, 2015	To dismantle criminal organizations conducing illegal mineral and gold trading in the north of Mato Grosso state	47	11
Corrida Do Ouro	Nov 6, 2015	To dismantle a criminal network of civilian and military police collaborating in mining and the sale of illegal gold in the municipality of Pontes e Lacerda	32	0
Madeira Limpa	July 24, 2015		41	3
Grand Canyon	Nov 23, 2015		14	5

(continued)

Operation	Date	Objective(s)	Search orders	Detentions
Cratons	Dec 8, 2015	To combat illegal extraction and sales of diamonds in the land of indigenous peoples of Cinta Larga in Rondônia	41	11
			659	415

Source División of environmental crimes of the police federal—DMAPH (2012)

Conclusion

The next two decades will be decisive for the Brazilian Amazon: If properly managed, the forest could be a permanent source of wealth for traditional communities, businesses, and workers; if not, the range of social, economic, and environmental problems resulting from illegal deforestation may drive the region toward a collapse of monumental repercussions. To avoid this fate, we must encourage greater awareness and participation by all branches of government and all sectors of civil society (universities, researchers, NGOs and citizen associations, indigenous inhabitants, professionals, farmers, and laborers). In particular, the state must heavily invest in its preventative and repressive apparatus, especially in the areas of intelligence and infrastructure, taking into account that environmental crimes are conducted by organized networks determined and capable of eluding the law and its enforcers. The best practices presented here illustrate the broad advances that Brazil has made in the past decade. But even with such progress, we must keep in mind that we are still just beginning to lay the path toward durable protection of the Brazilian Amazon. That path is long. Let's continue on it.

Notes

1. This office, equivalent to a US Cabinet Secretary, heads the President's relations with the Cabinet, Congress, and governors. This role makes the position the second most powerful in Brazil.
2. Water vapor produced by Amazonia's rain forest extends throughout South and Central America. Deforestation thus has a massive impact on water supply and hydroelectric energy (Fearnside 2005).

3. Data from the lumber sectors and production surveys by Imazon showed that state and federal tax deductions, just in 2009, approximated US $200 million (Adeodato et al. 2011). The amount of such tax evasion is greater than the annual budget of most enforcement agencies, including the Police Federal.
4. Deforestation in Brazil's Amazon is inextricably linked to violence in the region. Comparison of data from the National Institute of Spatial Research (INPE: Instituto Nacional de Pesquisas Espaciais) with the "Map of Violence of Brazilian Municipalities" by the Organization of Ibero-American States showed that, in 2005, of the 100 municipalities with the highest rates of deforestation, 61 also had Brazil's highest homicide rates (Waiselfisz 2016).
5. For information on all Federal Police operations between 2003 and 2016: http://www.dpf.gov.br/DCS/operacoes/indexop.html.
6. Data from international organizations show that the global timber industry moves nearly $100 billion annually, of which close to US $2.5 billon is connected solely to legal income of this sector in the Amazon.
7. Although the main objective of mining activity is not directly linked to extraction of forest products, such activity does cause serious impacts on the forest, including deforestation of immense areas. In particular, many of the chemical processes and products utilized in mining causes sedimentation of rivers, bringing serious consequences for human and animal health.
8. In Brazil, minerals are State property and extraction requires DNPM authorization, a federal agency of the Ministry of Mines and Energy (MME), as well as authorizations by environmental authorities overseeing forest management and mining in each area.
9. Brazilian Criminal Code article 75 limits incarceration of an offender sentenced in Brazil to 30 years.
10. The DOF (*Documento de Origem Florestal*) is the document authorizing the transport and commercialization of forestry products.
11. SIG is the Portuguese acronym for Geographic Information System (GIS).
12. Consultation in the *Mineiro Cadastre* verified that the applicant requested suspension of the processes of mineral exploitation in August 7, 2007 and that after that date the authorization was never granted to him again, reason why any exploitation after August 2007, in that area, is illegal.
13. This type of fraud in the issue of DOFs is common. To cover wood illegally extracted from public or protected areas, criminal organizations use DOFs of other legalized areas. But this fraud can be detected through incompatible data in DOFs—for example, if the total time the transport spent on the route is incompatible with the distance or the numbers of

the license plates of vehicles used do not exist, or, if they do, the vehicle is not licensed for the transport of wood (e.g., motorcycles).

14. The criminal organization had been able to "legalize" of an area of 146,000 hectares, and to obtain permits for logging on 21,000 hectares, which gave them about 1.4 million m^3 of wood or 56,000 trucks.
15. A "Special Operation" in the Brazilian Federal Police involves investigations using measures requiring judicial warrants, like telephonic tapping or exceptions to banking secrecy, and the search and seizure of evidence of criminal activity (e.g., cash, vehicles, jewelry).

Bibliography

Adeodato, S., M. Betiol, and M. Monzoni. 2011. *Madeira de Ponta a Ponta: o caminho desde a floresta até o consumo*. São Paulo: Fundação Getúlio Vargas.

AMERIPOL. Associação de Polícias das Américas. 2014. *Buenas Prácticas: un modelo de intervención policial efectivo y transcendente em América*. Bogotá: The Police Community of the Americas.

DMAPH. 2012. *Crimes Against the Flora Course*. Brasília: ANP (Academia Nacional de Polícia).

Fearnside, P.M. 2005. Desmatamento na Amazônia brasileira: história, índices e consequências. *Megadiversidade* 1 (4): 113–123.

GAGEO. 2016. *Relatórios, laudos, informações e dados produzidos pelo Grupo de Análise e Geointeligência da Divisão de Repressão aos Crimes contra o Meio Ambiente e Patrimônio Histórico da Polícia Federal (DMAPH/PF)*. Brasília: GAGEO.

Instituto Nacional de Pesquisas Espaciais (INPE). 2016. *Monitoring of the Brazilian Amazon Forest by Satellite 1999–2009*. São José dos Campos: INPE.

Mongabay. 2010. *Operation Jurupari and Illegal Logging in the Amazon*. https://news.mongabay.com/2010/07.

National Geospatial-Intelligence Agency (NGA). 2006. *National System for Geospatial Intelligence (GEOINT) Basic Doctrine*. Springfield, VA: National Geospatial-Intelligence Agency.

Perazzoni, Franco. 2012. Criminalidade Ambiental Organizada: o desmate ilegal em Mato Grosso. In *Fórum Nacional de Segurança Pública*.

Perazzoni, Franco. 2013. SIG, Amazônia e Polícia Federal: Geointeligência no combate ao desmatamento ilegal na Amazônia. Master in Science (Msc.) thesis, Universidade Nova de Lisboa, Lisbon.

Pereira, M., M. Panjer, T.S. Greenberg, and W.B. Magrath. 2011. *Justice for Forests: Improving Criminal Justice Efforts to Combat Illegal Logging*. Washington, DC: The World Bank.

United Nations Office on Drugs and Crime (UNODC). 2011. *Colombia: Coca Cultivation Survey*. Bogotá: UNODC.

Waiselfisz, J.J. 2016. *Mapa da Violência dos Municípios Brasileiros*. Brasília: OEI.

CHAPTER 3

Deforestation in the Bolivian Amazon: The Case of the El Choré Forest Reserve in Santa Cruz Department

Manlio Alberto Roca Zamora

Abstract This chapter systematically spells out the historical, political, and economic fuels of the massive deforestation in Bolivia's Amazon, focusing on enforcement in Santa Cruz, Bolivia's largest and most eco-diverse department (state), which for the past 10 years has also been the country's primary battleground of political control as well as its engine of export-led economic growth. With a case study of a major Forest Reserve in Santa Cruz, this chapter examines the efforts to create a functioning system of enforcement, without adequate authority or autonomy, under these enormous ecological and economic pressures.

Keywords Amazon · Environment · Nature reserves

Translated by Alix Shand.

M.A.R. Zamora (✉)
Department of Santa Cruz, Santa Cruz de la Sierra, Bolivia

M. Ungar (ed.), *The 21st Century Fight for the Amazon*,
https://doi.org/10.1007/978-3-319-56552-1_3

The Context

Bolivia is better known as an Andean country rather than for its far-reaching lowlands that cover over 65% of its territory, including the Amazon, the Chiquitano dry forest, and part of the Chaco Boreal. All of the nature inventories conducted in the past few decades affirm that Bolivia is a megadiverse country. Its tropical region at 9° to 23° south, the Andes soaring from to 15,000 feet, and the vast diversity of habitats together contain an incredible wealth of species. Even though Bolivia is home to around 3.5% of the world's forests, 30–40% of the world's biodiversity can be found here. It is estimated that there are around 20,000 species of vascular plants, around 390 species of mammals, 1422 species of birds *(Asociación Armonia BirdLife database)*, 306 species of reptiles, 255 species of amphibians, and over 970 species of fish on Bolivian territory. This megadiversity of species is in sharp contrast with our limited knowledge on some of the taxonomic groups (e.g., invertebrates in general).

Three criteria can be used to describe the Bolivian Amazon region:

(a) Hydrographic: comprising all of the basins that drain into the Amazon basin covering 716,370 km^2 or approximately 65% of the country.
(b) Biogeographic: corresponding to the Amazon Biome, defined as an area predominantly covered by dense inundated tropical forest including relatively small areas of different types of vegetation, such as savannas, floodplain forests, meadows, marshland, and palm forests. The Bolivian Amazon covers a surface area of 475,278 km^2, almost 43% of the country.
(c) Political: Bolivia is divided into nine departments (states), which themselves are divided into provinces. As stipulated in the Constitution of the Plurinational State of Bolivia, enacted on February 7, 2009 under President Evo Morales, the "Amazon Region" is comprised of the Department of Pando, the provinces of Ballivián and Vaca Diez in the Department of Beni, and the province of Iturralde in the Department of La Paz. Together, its surface area is 141,923 km^2, which is 13% of national territory. Left out are large parts of Beni, the Tropical east of Cochabamba, and the north of Santa Cruz. The political determination of what constitutes the Amazon seems to have been made on a whim,

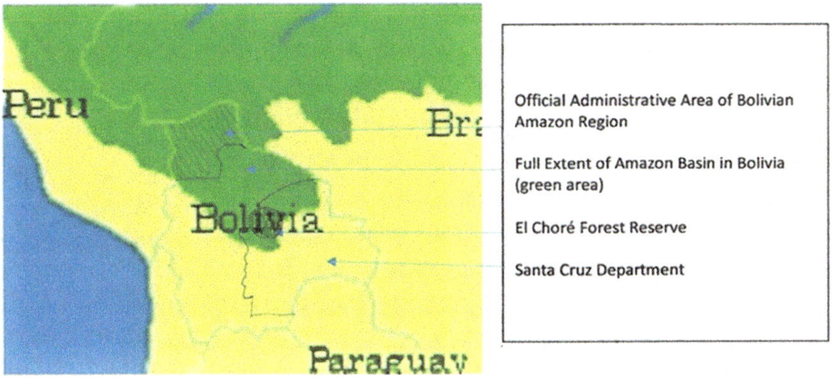

Fig. 3.1 The Bolivian Amazon: Full Basin, Biome, and Political-Administrative. *Source* Created by Mark Ungar

> resulting in a highly truncated and inaccurate political-administrative boundary (Fig. 3.1).

International and conservation organizations, such as the ACTO, IUCN, WWF, WCS, and CI, use technical and scientific criteria that are more accurate and extensive than the biased vision of the current politicians that decided to redraw the borders of the Bolivian Amazon.

The following maps show the difference between the two definitions of the Bolivian Amazon region.

The Background: The Bohan Plan and the March to the East of Bolivia

At the end of the Chaco War against Paraguay in June 1935, Bolivia was in ruins and in urgent need of rebuilding. And this is when a US mission, led by Melvin Bohan, arrived in Bolivia to design a huge project to drive development, which eventually became known as the Bohan Plan (1941). The members of the mission concluded that Bolivia could no longer rely on mining to prop up its economy due to fluctuating international mineral prices, and that the way ahead led to the east, to Santa Cruz, which had been largely ignored by politicians. The region, according to the report, would be the hub of development in the following decades. A major road network, increased agro-industrial production, credit facilities, agricultural expansion and development, and gas and oil exploration were the main components of the Bohan Plan. The proposal focused on Santa Cruz as the region with the greatest potential for economic growth. The Bohan Plan proposed creating a road network, with the paved Cochabamba—Santa Cruz highway as the centerpiece; improving farming and cattle ranching; instituting agro-industrial plants and credit agencies (banks) developing existing oilfields and exploring new reserves; developing markets; and other strategies for expanding land for agriculture and cattle ranching.

In 1942, as a result of the Bohan Plan, the government set up the Bolivian Development Corporation (CBF) to drive the country's agro-industrial development, which included building sugar refineries. It also set up the Banco Agrícola to provide credit to the budding agro-industrial sector and, in 1943, started work on building the paved Cochabamba-Santa Cruz highway, which would be the first paved road in Bolivia. Most Bolivians, and of course the economic transactions among them, were based in the highlands and inter-Andean valleys, and were made up primarily of traditional mining and small-scale family farming. In the face of the country's dependence on mineral exports, the government sought to diversify the economy and curb food imports by increasing food security, and start industrializing the country. The solution to the problem of foreign exchange reserves and a population that barely scratched out a living from the earth in the highlands and valleys lay in reaching the fertile, uninhabited lowlands of the Bolivian east to produce the staples that Bolivia had been importing up until then. This policy of populating "empty" space, as the region was known as at the time, was called the "March to the East." The process that mobilized the transference was called "colonization," fueled by a policy in which state-owned lands were given to groups of small-scale farmers from the highlands. Under this program, large numbers of Aymara and Quechua families, called colonizers, moved to the lowlands of Santa Cruz. The March to the East was a defining element of the Bohan Plan. Many Bolivians took the opportunity to get ahead in a country whose mineral reserves were fast dwindling.

Deforestation in the Bolivian Amazon: The Case of Santa Cruz

As described above, the policy to occupy the east of the country, mainly the Department of Santa Cruz, spurred the internal migration of Quechua and Aymara families, who went from living and farming on small unproductive plots of land in the highlands to medium and large-scale lands in the Bolivian forest. This completely different environment—almost like night and day in terms of climate—added to their culture shock and upended their idea of how an agricultural holding should be formed and operated. Based on their traditional practices, which the only ones they knew for centuries, the colonizers cleared small family-size plots in the forest, planted crops, and built dwellings to take refuge from the unknown and dangerous "green hell," as they called the rainforest. And thus began the slashing and burning of thousands of plots in the so-called *nucleos* of colonization. This practice then became one of the main drivers of deforestation in Eastern Bolivia, propped up by each successive government. Military juntas, the leftist populists in the 1950s, and the Western-oriented neoliberal Presidents that ruled after the transition to democracy in the 1980s, all supported the Plan and, even when recognizing or regretting its impacts, ultimately overlooked its damages on the environment that have little by little affected large areas of Santa Cruz's tropical forest.

A few years later, in the 1970s, the emerging agriculture and agro-industry, originally based on sugarcane in the north of the department, expanded by including two more extensive crops—cotton and soybeans. The former had a relatively short lifespan because of poor decisions regarding sales to other countries, whose governments then closed the door to Bolivian cotton. Soybeans, however, became the country's most important oilseed crop due to the high price it fetches and a growing international demand, especially from the Asian market. In the 1980s, soybean crops were joined by other important industrial crops, such as corn, sunflower, and sorghum. The huge Brazilian agricultural companies concentrated along the border with Bolivia also brought in new technologies and ideas of how land should be worked. The fertile soil in this part of Bolivia, low labor costs, and the fact that the people of Santa Cruz have always been open to new initiatives, opened up and consolidated the agro-industry that defined the Department of Santa Cruz as the country's epicenter of development. Its capital city, Santa

Cruz de la Sierra, became one of the fastest-growing cities in Latin America.

The new millennium thus saw Santa Cruz become home to large areas of agro-industrial crops, foreign agriculture companies, colonies of Mennonite and Russian farmers settling on large estates in the Bolivian east (it has been calculated that the Mennonite colonies own over 500,000 hectares in the Department of Santa Cruz alone). Alongside agricultural activity in the east, cattle ranching also went through some significant changes in terms of pasture areas, yields, husbandry, and genetic improvement. Cattle ranching is an activity that is constantly scaling up and requires new land for the cattle to graze on, which in turn leads to intense growth of the herd. And so it became a second major driver of deforestation which, along with the first, has had a series of social and environmental impacts on the Amazon region of Santa Cruz. When quantifying the loss of tropical forests, 1000 small-scale farmers clearing and burning 10 hectares a year each, or ten companies with 1000 hectares each doing the same, wreak the same destruction on the Amazon forests, and has given Bolivia one of the highest per capita rates of greenhouse gas (GHG) emissions in the world—just by burning the biomass in deforested areas (See Fig. 3.2).

Data from the World Wild Fund (WWF) for the last 50 years show that the seemingly infinite Amazon has lost at least 17% of its forest cover, and that many of endemic species are being brought to the edge of extinction. Economic transformation in the Amazon concentrating on converting and destroying the natural habitat is speeding up, but so too, as mentioned in the introduction, is understanding of how great a role the Amazon plays in regulating local and global weather.

In Bolivia, Santa Cruz is by far the largest of the country's departments, with a surface area of 370,621 square kilometers, covering 33.7% of national territory. Its location provides for very heterogeneous physiographical and environmental features that can be seen in the biological diversity mentioned at the beginning of this chapter, and which makes Bolivia one of the world's 17 megadiverse countries. But the department now also has one of the biggest populations, with almost 2.7 million inhabitants (according to the Census, 2012), many of whom were drawn by the fact that it has the country's fastest growing regional economy. That growth is centered on highly integrated productions that are taking over huge spaces for farming, petroleum exploitation, mining, and industry (see Vides-Almonacid et al. 2015).

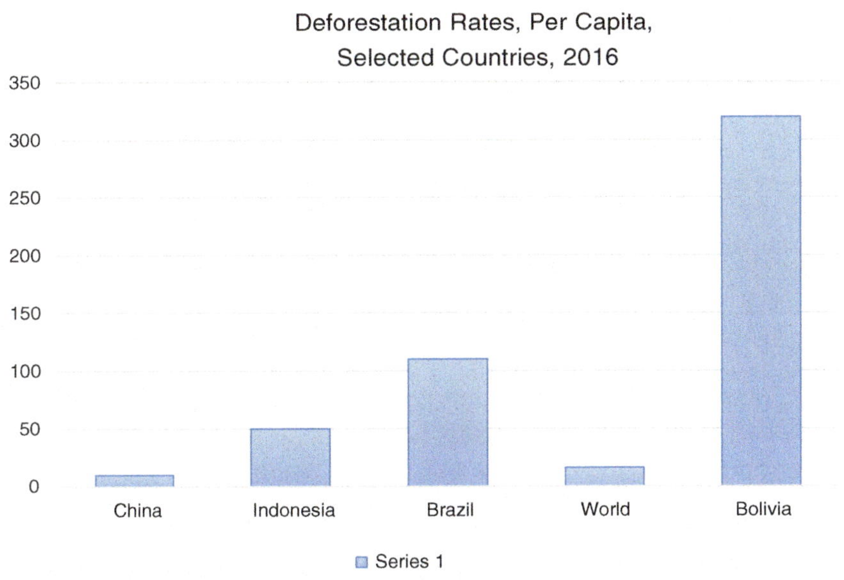

Fig. 3.2 Per capita deforestation rate. *Source* Created by Mark Ungar

The main threat to the biodiversity in the Department of Santa Cruz is deforestation and other forms of natural habitat transformation from the continuing expansion of the agricultural frontier. This activity has unleashed a series of stressors that directly or indirectly affect biodiversity and ecosystem stability: building roads, new settlements, hunting, illegal logging, and forest fires, among other destructive activities. Removing forest cover leads directly to habitat loss for plant and animal species, exhaustion of forest resources, less ability of the forest to capture carbon, the accumulation of greenhouse gases, the reduction of the biomass, erosion and compacting of the earth, sedimentation of rivers, and, ultimately, climate change. Even in areas with no direct loss of the forest canopy, the forests are experiencing accelerated degradation as seen through their original state becoming impoverished and changed. The environmental impact does not only affect the environmental goods and services provided by the forests, but also the soil productivity as a result of the intensity of mono-cropping and the lack of proper management, in many cases leads to an increased rate of degradation. The felling of the trees that cover the water basins leads to declines in water quality and changes in the water regime, increasing the risk of flooding and speeding up the processes of erosion, both by water and wind.

According to data from the Atlas Socioambiental de Tierras Bajas y Yungas de Bolivia (FAN 2015), 7.4% of the Bolivian Lowlands and Yungas have been deforested in the last 40 years (5.7 million hectares in total) at an average annual rate of over 190,000 ha/year in the last decade. Between 1970 and 2000 almost 3.1 million hectares of forest were lost, and from 2000 to 2013 a total of 2.6 million hectares were lost. Separation of these figures by department shows that Santa Cruz has suffered the greatest loss of forest, representing 78% of all the deforested area in Bolivia: Accumulated deforestation in the Department of Santa Cruz is over 4.4 million hectares and represents 13% of the department's surface area, of which 56% (~2.5 million hectares) was lost during the period 1970–2000, and 44% (~2 million hectares) was deforested between 2000 and 2013 Table 3.1.

Santa Cruz is the department whose ecosystems have suffered the most from land conversion. Promotion and subsidies for its primary field crops, such as oilseeds, along with increasing cattle ranching, which requires huge surfaces for pasturelands, do not account for their combined social and environmental cost, which all too often are unknown or underestimated. Since 2000, in particular, cattle ranching has been the primary driver of deforestation in Bolivia; it is estimated that between 2005 and 2010 it was responsible for 60% of land clearing. Cattle ranching is affected by the climate much less than is agriculture. The main factors for its expansion are influence from Brazil, foreign investment, and interest from foreign markets. Studies show that in the future this use of the land will be a major threat to Bolivian forests. In the east of Bolivia, environmental concerns clash with agricultural interests, since the economic "value" of forested land is not registered if not being used as farmland.

Given the lack of concern or direction from successive federal governments, local governments took up the slack by designing sustainable public policies for development focusing on regional needs and meeting the challenge of making preservation of the Department's biodiversity compatible with productive development. One of the first manifestations of this approach was the 1995 Santa Cruz Land Use Plan (PLUS), set in motion with help from the German Development Bank (KfW), which is part of that country's GIZ, the Bank of Financial Cooperation. The PLUS was a planning tool for sustainable regional development used to

Table 3.1 Forest loss in Santa Cruz by period

DDepartment	*Area of Lowlands and Yungas (ha)*	*Forest loss by period to 2013 (ha)*							*Total surface area deforested (ha)*	*Area deforested by 2013 (%)*
		Before 1976	*1976–1986*	*1986–1991*	*1991–2000*	*2000–2005*	*2005–2010*	*2010–2013*		
Santa Cruz	35,221,198	556,620	258,363	452,543	1,185,234	719,368	669,534	570,219	4,411,881	13

Table 3.2 Most common land-use categories in Santa Cruz

Plus categories	*Surface area (ha)*	%
Intensive farming	2,982,885	8
Extensive farming	6,247,197	17
Agrosilvopastoral	3,496,117	10
Forestry	9,062,737	25
Restricted use	3,148,220	9
Protected areas	11,206,277	31

Source Based on data from PLUS

formulate a spatially explicit set of recommendations on land-use based on criteria such as soil capacity, biodiversity, topography, hydrography, and climate, and seen in Table 3.2.

In 2012, the Autonomous Departmental Government of Santa Cruz, aware of the damage of the irrational use of the land and the systematic depredation of renewable natural resources, decided to step up its land planning process by developing a Departmental Land Management Plan (PDOT), taking into account the department's strengths and weaknesses, and the social and economic initiatives needed to make Santa Cruz attractive economically yet sustainable socially and environmentally. In the PDOT, land-use must be carefully planned and approved, with new restrictions on harmful agricultural activity.

While helpful, restrictions of this regulatory framework have been broadly disregarded, leading to the degradation of significant tracts of forest land and restricted-use land, doing irreversible damage to the areas that were designated for purposes other than agriculture. So, the question remains: How is it possible that in Bolivia and in Santa Cruz, where there is a fairly decent technical and legal framework for regulating land-use and environmental impact, there are so many illegal activities and nothing is being done to stop and punish those responsible? The answer is complicated, involving a range of social-political actors and scenarios that are particular to Bolivia.

Case Study of the Environmental Threat in Santa Cruz: The El Choré Forest Reserve

A key example of illegal deforestation in Santa Cruz is the case of the El Choré Forest Reserve, located between the Ichilo and Yapacany rivers, which are the main tributaries of the upper basin of the River Mamoré.

The Mamoré joins with other rivers also originating in Bolivia to flow into the River Madeira, which provides most of the sediment that feeds the diversely rich aquatic life of the great Amazon River itself.

The El Choré Forest Reserve is the last patch of Amazon forest in the north of Santa Cruz that helps to regulate the weather in the northern provinces of the department, and is an area of concentrated and extensive biodiversity. Listed as one of Bolivia's most productive forests, the Reserve was created on August 3, 1966 with a surface area of around 900,000 hectares, which later was increased to 1.08 million hectares because of its enormous wealth of forest resources. It was the first forest reserve to be created in Bolivia with the aim of sustainably extracting forest resources, but since 1970 the southeastern part of the reserve has been plagued by illegal settlements of colonizers.

In September 1995, when the Santa Cruz PLUS was enacted, the concept of a forest reserve changed. A new category—the "B2" Sustainable Management Forest—was introduced, which covers 784,538 ha. An Agroforestry Unit was also formed, covering 223,962 ha for small-scale farmers to settle on, known as the "unaffected area" of El Choré. Regulations for these new areas established the following parameters:

Prohibit the entry of new colonizers into the zone. Limit more colonization for agricultural purposes. Substitute existing agricultural systems with pastoral systems that conserve the soil. Prohibit mechanized land clearing. Prohibit mechanized farming. Implement the animal health program. Protect the wildlife. Strengthen and coordinate forest research, outreach, and control institutions.

One of the main aims of restricting access to the unaffected area was to reduce the pressure on the forest that was still intact within the reserve. Unfortunately, this measure could not stop the assault on the rest of the reserve; in fact, more settlements were established and conflicts between the loggers and the *interculturals* (as the highland colonizers are now known) quickly multiplied. The conflicts over land and the forest peaked in the mid-1980s when the armed forces were called into prevent new settlements. In 1985, a truce was called and a "red line" was drawn to divide the land destined for economic activities from the land occupied by the communities of *interculturals* in the southern end of the reserve. In addition, the farmers agreed to avoid colonizing areas outside the reserve's borders. The inhabitants of the village of Santa Rosa, the capital of one of the three municipalities that make up the forest reserve, have traditionally always been ranchers and loggers, and greatly resent the flood of *interculturals*, believing that they had come to take their lands and forests. This was a reason why the *interculturals* were hardly involved in local politics, which eventually changed when the first term of the neoliberal government of Gonzalo Sánchez de Lozada (1993–1997) greatly expanded citizen participation.

The Popular Participation Law came into effect on April 20, 1994 and is considered to be the most significant measure of administrative and financial decentralization in Bolivia over the last 50 years. Realizing that it could not run the country from the capital of La Paz or even from the nine department capitals, the Sánchez de Lozada administration believed that transferring power, resources, and authorities to the municipal governments would trigger local creative capacities and build development responsibly. Popular participation was the Bolivian contribution to the different decentralization processes in Latin America and is still a valid administration model due to its social oversight, development management and municipal strengthening components. In a very short time, local leaders across the country got on board with this move to popular participation. But it has had different impacts in different localities. The administration of Santa Rosa, El Choré's primary municipality, has changed radically as the migration of interculturals from the highlands continues. These migrants have displaced the local people and traditional power-holder groups, and have even risen into public office, becoming mayors, counselors, and even members of the departmental assembly in the Santa Cruz parliament.

Those that have suffered the greatest pressure, however, are the indigenous peoples of the Amazon forests. In 1992, for example, the World Vision Church relocated the remaining members of the Yuki people to Chimoré, a small village in the Tropic of Cochabamba, taking them out of their natural habitat. Only the Yuracaré and Mojeño peoples remain in small communities along the banks of the Ichilo River. Each group of people have a local leader known as *cacique*, *gran cacique*, or *Buitá Matata*.

More immediately, how can the problem of illegal settlements in El Choré be solved? As described above, the different public and private actors involved in the reserve form a tangled web, making it difficult to find a solution that is environmentally sustainable, socially viable, and politically harmonious. The recent publication "Problems Facing the Forest Reserves in the Department of Santa Cruz" states that almost half a century after illegal settlements sprang up in the El Choré Forest

Reserve, the central government needs to muster the political will to put aside private, sector, and political interests. But also required is coordination among the three levels of government—national, departmental, and municipal. It also requires that the interculturals living in the forest be willing to engage in dialogue to look for solutions, with no restrictions or pressure from any side. The aim must be to counterbalance the power of the interculturals, who are implicitly supported by the central government's inertia and the complicity of certain politicians (deputies, mayors, and councilors) who support the reversal of the forest reserve status due to political interests that border on the illegal. It is also important to consider setting up a body that represents civil society that works on defending the land, mobilizes, raises awareness, and helps make decisions about the situation of the interculturals living in the area.

Although the conflicts and human settlements in the El Choré Forest Reserve date back to its foundation, they have worsened with time. There have been numerous attempts to conserve and protect the reserve, but it is clear that the government is the main protagonist and does not show any interest or commitment to solving the problem. Moreover, the agricultural frontier of the Chapare (the region in the neighboring Department of Cochabamba where illegal coca is grown, destined for narcotics production) is encroaching into the El Choré Forest Reserve. As a result, any legislative bill on the reserve is a possibly fatal one for El Choré if not accompanied by the government's commitment to protect it. Departmental and municipal governments simply cannot take the lead if the central government owns the land, decrees the laws and mobilizes the resources for its own interests. Even worse, a proposal was made at the end of 2016 that may sound the death knell for the El Choré Forest Reserve. A group of politicians linked to the current ruling party is lobbying to revert a huge area of illegal settlements (approximately 450,000 hectares) so that these state-owned lands that make up the reserve can be handed over to agrarian unions linked to the ruling party.

The problem in this case goes beyond just the forest reserves, since it also involves two protected areas, private properties, and unplanned urban settlements. It is therefore critical that decisions be made quickly involving municipal, departmental, and national authorities, as the situation is turning critical. It will be more difficult for the municipal governments to deal with the social demand for new settlements (given that they are illegal by nature) that are expanding rapidly along with a population that will demand health, education, transport, and other public services.

This is the unfortunate predicament of the El Choré Forest Reserve, the richest in Santa Cruz and one of the most important in the Bolivian Amazon. As long as no solution is found, deforestation will persist, illegal settlers will continue to practice slash and burn, large forest fires will burn, and biodiversity will contract—all putting at risk the loss of this ecosystem contributes to the country, such as evapotranspiration, protection of aquifers, protection against flooding, windbreaks, control of erosion, capturing carbon, being among the most important.

To sum up, El Choré is one small illustration that the legal framework of the Plurinational State of Bolivia—from the new Constitution to all the laws protecting and our natural heritage and the passionate speeches about protecting Mother Earth made by our governors at international events—are all smoke and mirrors designed to hide Bolivia's dire environmental reality.

References

Fundación Amigos de la Naturaleza. 2015. *Atlas Socioambiental de las Tierras Bajas y Yungas de Bolivia*. Santa Cruz de la Sierra: FAN.

Vides-Almonacid, R., N. Pacheco, D. Anivarro, and D. Rumiz. 2015. *Problemática de las Reservas Forestales en el Departamento de Santa Cruz*. Santa Cruz de la Sierra: Gobierno Departamental de Santa Cruz.

CHAPTER 4

Peru: A Legal Enforcement Model for the Amazon

Hugo R. Gómez Apac, María Antonieta Merino de Taboada and Milagros Granados Mandujano

Abstract Written by the directors of one of the Amazon Basin's strongest enforcement bodies, this chapter describes, in great detail, how a national agency puts together a nationwide system of environmental enforcement able to reach into every activity and region. It also shows how that system grapples with the country's vast Amazon region amid clamors for economic development, tensions between the central government and local populations, the uncertainty of national politics, and other formidable roadblocks for what could be an international model of ecological enforcement.

Translated by María Antonieta Merino de Taboada.

H.R. Gómez Apac (✉)
Universidad Nacional Mayor de San Marcos, Lima, Peru

M.A.M. de Taboada
Universidad Peruana de Ciencias Aplicadas, Lima, Peru

M.G. Mandujano
Servicio Nacional de Certificación Ambiental Para Las Inversiones Sostenibles, Lima, Peru

M. Ungar (ed.), *The 21st Century Fight for the Amazon*,
https://doi.org/10.1007/978-3-319-56552-1_4

Keywords Amazon · The Judicial Branch · Environment

Introduction

The Peruvian Amazon extends over 782,880.55 km^2 of the Amazon basin, occupying 13.05% of its continental reach. It is home to approximately 38,020 species of plants, mammals, birds, reptiles, and amphibians (Programa de las Naciones Unidas para el Medio Ambiente (PNUMA) et al. 2009, p. 112), making Peru one of the planet's few megadiverse countries. The region also provides critical services, such as water cycle regulation, capture of carbon dioxide (CO_2), and a supply of forest products. That combination of preservation and provisions has given Peru the challenge of taking advantage of its natural resources while also developing a sustainable development for future generations. Such a difficult challenge will require joint efforts of public and private actors—citizens, companies, and public authorities—to and comply with established laws and implement innovative actions. To direct such efforts, in 2008, Peru created the Agency for Environmental Assessment and Enforcement (OEFA, for its Spanish acronym), whose primary duty is control of economic activities that impact the environment as well as the performance of other public entities in charge of environmental control. This chapter assesses OEFA's performance in those roles. By describing how even this well-prepared centralized authority struggles to manage such enforcement, it provides lessons for the wider Amazon region.

Peru's Legal Foundation of Environmental Protection in the Amazon

The legal framework to protect the Peruvian Amazon is based on international and national law.

International treaties and declarations: Based on the recognition of States' obligation to protect the environment in the 1966 United Nations International Covenant on Economic, Social and Cultural Rights, Amazon Basin countries began to forge conventions for the conservation and sustainable use of its resources. In 1978, Peru and the other eight Basin countries signed the Amazon Cooperation Treaty to promote joint actions for harmonious development of the Amazon based on the preservation, conservation, and rational use of its natural resources. This treaty then gave rise to the Amazon Cooperation Treaty Organization (OTCA) in 1998. Subsequently, States parties to the United Nations adopted other international instruments of environmental protection, such as the 1982 World Charter for Nature, the 1992 Rio Declaration on Environment and Development, and the 1992 Convention on Biological Diversity. These advances all boosted promotion of natural resource conservation in territories like the Amazon, which are critically important for global attainment of sustainable development.

The Constitution: Article 2 of the Peruvian Constitution recognizes the fundamental right of everyone to enjoy a balanced and adequate environment. To support that right, the Constitutional Court affirms that the State has unavoidable obligations to maintain environmental assets, and that such obligations extends to individuals. The Constitution also has articles (66–69) regarding the protection of the environment and the sustainable use of natural resources, which stipulate the duties to conserve biodiversity and to regulate the use of natural resources, among others. Article 69, in particular, obliges the State to promote "the sustainable development of the Amazon with appropriate legislation."

Legal framework: Peru has steadily built a system of law to protect its Amazonia. The centerpiece of that protection is the General Environmental Law (Law 28611), which states that such protection must be part of the basic guidelines of any public policy adopted by the State, including the National Environmental Policy. In compliance, the 2009 National Environmental Policy made the Amazon's sustainable development a key objective, to be realized through several approaches: Promotion of research and use of clean technologies in mining and energy activities in the region; Recovery of areas degraded by illegal mining; Protection of Amazonian primary forests and gradual reduction of logging, slash, and burn for agricultural purposes; development of productive alternatives for forest use, such as ecotourism; Promotion of conservation and sustainable use that accounts for its variability, complexity, fragility, and geostrategic location. A central measure implemented to comply with the National Environmental Policy is the establishment, as discussed in the chapter on Ecuador, of Protected Natural Areas (hereafter PNAs)[1] throughout that region. As defined in Law 26834, PNAs constitute continental and marine spaces of the national territory in order to conserve biological diversity and other associated values of cultural, scenic, and scientific interest, as well as to contribute to the sustainable development of the country. Currently, due to the richness of the biological diversity in the Amazon region and the importance of its conservation, the State has established 36 PNAs, which fall into the following categories: ten National Parks, three National Sanctuaries, one Historical Sanctuary, five National Reserves, ten Community Reserves, three Protected Forests, and four Reserved Areas.[2] The protection of these natural areas is vital for the conservation of the Peruvian Amazon.

OEFA and Environmental Protection

To fulfill its duty to protect the right to a healthy environment, the State is obliged to adopt actions aimed at the conservation, prevention, repair, or compensation for changes to the environment. In 2004, Peru established a regulatory framework called the Framework Law of the National Environmental Management System (Law 28245) (SNGA for its Spanish acronym) to guide, integrate, coordinate, oversee, evaluate, and ensure the implementation of policies, plans, programs, and actions for the protection of the environment and the conservation of natural resources. SNGA is comprised of a set of agencies, led by the National System of Environmental Impact Assessment (SEIA for its Spanish acronym), the National System of Natural Areas Protected by the State (Sinanpe: El Sistema Nacional de Áreas Naturales Protegidas por el Estado), The National System of Environmental Information (Sinia: Sistema Nacional de Información Ambiental), and the National System of Environmental Assessment and Control (Sinefa: Sistema Nacional de Evaluación y Fiscalización Ambiental). Sinefa was created in 2009 through the National System of Environmental Assessment and Control Law (Law 29325). It comprises the Ministry of the Environment (hereafter *Minam*), OEFA, and the Environmental Inspection Entities (hereafter EIEs) at the national, regional, or local level. Sinefa aims to ensure compliance with environmental legislation, which includes verification of the oversight in environmental matters by various State entities. As both a governing and specialized technical agency affiliated with Minam, OEFA carries out environmental control actions at two levels. First, it supervises the environmental obligations of companies in the areas of mining (medium and large-sized), energy (hydrocarbons and electricity), fisheries (industrial fishing and aquaculture), and industry (beer, paper, cement, tannery, biofuel, among others). As Sinefa's governing body, in addition, it supervises the performance of the EIEs at the national, regional, and local levels and issues regulations for them.

OEFA's environmental regulation centers on *ex post* control of companies' activities, supervising them once the economic operation, including exploration, has begun. This control process includes a diagnosis of the environmental quality of the area of the proposed project or activity, the oversight of environmental obligations and, if non-compliance is detected, processing the respective sanctioning procedures, in which sanctions can be imposed and precautionary and corrective measures can be issued. The process also includes granting of incentives to companies for good environmental practices (Gómez Apac and Granados Mandujano 2015, p. 78). Altogether, OEFA's control centers on six main roles:

Environmental assessment: This function includes monitoring of environmental quality and development of comprehensive environmental assessments to do so. Such monitoring centers on samplings taken at different points in order to determine the state of environmental components (water, air, soil, flora, and fauna) in each area's ecosystem (Organismo de Evaluación y Fiscalización Ambiental (OEFA) 2015, p. 92). The second has greater complexity, since it consists of generating environmental diagnoses that involve the identification of possible sources of contamination and, in some cases, the determination of levels of risk to the environment in extensive areas, such as watersheds, exploration lots, reservoirs, aquaculture zones, energy nodes, and bays. The results obtained from these actions may lead to subsequent oversight actions in order to verify whether the companies opreating in the area comply or not with their environmental obligations and whether the modification or alteration of an environmental component is due to those companies' economic activities.

Oversight: OEFA's oversight function centers on follow-up and verification of the activities of companies in order to ensure a good environmental performance and compliance with their environmental obligations, as specified in: (i) environmental regulations, (ii) environmental management instruments, (iii) administrative orders issued by OEFA, and (iv) other sources of environmental obligations (Gómez Apac and Granados Mandujano 2015, p. 82). The oversight function is aimed at preventing environmental damage and promoting voluntary rectification of alleged breaches. Since 2013, OEFA has been emphasizing preventive and corrective environmental controls that promote such voluntary correction. Law No. 30011 of 2003 first introduced this approach by issuing regulations stipulating that if a company rectifies minor breaches—involving either actual or potential harm—then it would not be subject to sanctions. In 2015, a new regulation extended the scope of this benefit to breaches qualified as "moderate," which involve potential damage to flora and fauna. Through a preventive and corrective approach, these regulations aim to avoid environmental damage and promoting voluntary rectification of violations. Such an approach is seen as maximizing the best outcome for both for companies and the state.

Administration: On the other hand, under the oversight function, OEFA is empowered to issue administrative orders, including preventive measures, mandatory mandates, and requirements for updating environmental management instruments. Administrative orders are issued by OEFA bodies to companies to comply with one or more of three measures. The first is Mandates: Aimed to order companies the elaboration or generation of relevant information or documentation to ensure the effectiveness of environmental control and which go beyond simple information requirements. This type of measure includes the preparation of technical environmental studies, monitoring programs, and others of a similar nature that generate information on environmental performance. A second required measure is for updating environmental management instruments, issued when negative environmental impacts are detected and differ significantly from those stated in the environmental certification. Under this measure, the company must request an update of its environmental management instrument to include the actual impacts of its activity. A third set of measures are preventive and exceptional measures issued when there is an imminent danger or high risk of serious damage to the environment, natural resources, or public health, as well as to mitigate degradation or damage to the environment. These measures are not necessarily related to alleged breaches, and are based on the principle of prevention in the General Environmental Law (Law 28611).

Enforcement and sanctioning function: OEFA can investigate any suspected breaches of environmental obligations. If a breach is demonstrated, sanctions or corrective measures can be imposed. There are four such measures, all centered on the principles of proportionality and reasonableness: (i) *Adequacy*: The company must adapt its activities to specific standards, often through mandatory training courses, to ensure the mitigation of possible harmful effects on the environment or human health; (ii) *Standstill*: To stop or neutralize an activity causing environmental damage, through measures such as confiscation of goods or the temporary or definitive closure of establishments; (iii) *Restoration*: To restore, rehabilitate, or repair the altered situation, such as to restore a faunal population decimated by environmental contamination; and (iv) *Environmental compensation*: For an environment that cannot be restored, these measures include compensatory reforestation on nearby land with similar development possibilities.

Incentives: Environmental control also includes granting incentives to companies in order to encourage good practices (Lozano 2010, p. 595). This system is run through OEFA's Register of Good Environmental Practices, in which it records companies that fully comply with both state environmental regulations and their own commitments. Registered companies can also earn incentives if they carry out practices beyond those required by state regulations, their own commitments or orders issued by authorities. To encourage such over-compliance, OEFA has a range of honorific and economic incentives. Honorific incentives include (i) the incorporation into the Environmental Excellence Ranking—REAL, (ii) the annual Qumir Rapi award (green leaf), and (iii) the annual Qumir Kawsay award (for "green living"). Economic incentives include a Certificate of Discount on Fines, which represents a value in taxation units (UIT) that is transferable and is valid for 4 years.

The EIEs: But perhaps OEFA's biggest and most difficult oversight role is Control of Environmental Inspection Entities (EIE). OEFA is authorized to supervise, issues regulations for, and verify performance of the Environmental Inspection Entities (EIEs), which conduct most environmental controls on the muliple levels of Peruvian governance, centered on a federal government and the country's 25 regions (analogous to US states), each of which is subdivided into provinces (provincias), which themselves are composed of districts. At the national level are ministries and specialized agencies with environmental control functions. For example, the Ministry of Production oversees fisheries. Regional governments, for their part, have environmental oversight in the mining, agriculture, health, fisheries, and tourism subsectors, as well as small-scale and artisanal mining. EIEs at the provincial and district municipalities, finally, also have environmental control functions, such as over control of solid waste management. OEFA carries out on-site and documentary oversight of all these EIE duties, often with technical assistance and training to ensure that they are carried out. When it discovers a breach of duties, OEFA reports it to the Peruvian General Comptroller to further determine legal responsibility. OEFA may also file criminal complaints through its Public Prosecutor's Office, which covers two main areas: actions or omissions of public officials or EIE personnel that may lead to criminal liability; and actions or omissions by natural or legal persons supervised by the EIEs that may cause a serious risk or damage to the environment or human health, and thus may constitute environmental crimes.

Principal Problems in the Amazon: OEFA's Actions and EIE Oversight

OEFA oversees all activities affecting the environment, with a focus on mining, forestry, and solid waste. Mining activities, a key issue, are classified into four types, based on the magnitude of their environmental impacts: artisanal, small-scale, medium-sized, and large-sized (Mining Law article 91). Depending on the category in which their activity is classified, miners must have certain types of environmental management instruments (IGA for its Spanish acronym). Artisanal and small-scale mining producers (SMP) require an Environmental Impact Statement (EIS) or a Semi-detailed Environmental Impact Assessment (EIA-sd); medium- and large-sized producers need an EIA-sd or Detailed (EIA-d). As shown in Table 4.1, control over these instruments corresponds to different public authorities, depending on their levels of expertise and capacity. Since OEFA has greater technical capacity and resources to oversee activities with greater environmental impact—medium-sized and large-sized mining in particular—it supervises them. Regional Governments (GORE) oversee small-scale and artisanal mining (Ministry of Environment 2016, p. 32).

Table 4.1 Entities responsible for the environmental control of mining activities

Type	*Concession size*	*Installed production capacity*	*Responsible state authority*
Large-sized mining	A limit has not been stablished	More than 5000 TMD	OEFA
Medium-sized mining	A limit has not been established	Between 350 and 5000 TMD	OEFA
Small-scale mining	up to 2000 Ha	up to 350 TMD	Regional Government
Artisanal mining	up to 1000 Ha	up to 25 TMD	Regional Government

Source Created by authors

Mining is illegal when carried out without proper titles or in prohibited areas. It is by far one of the biggest sources of damage not just to biodiversity and the forests, but to the health of the people who live in them, mainly from absorption of the mercury, lead, and arsenic commonly used in this activity. Those substances also contaminate bodies of water—rivers, lagoons, and lakes. A study by the Carnegie Institute determined that 60% of fish consumed in Puerto Maldonado (an area with high rates of illegal mining) had mercury levels far above those "allowed" by the World Health Organization (WHO), and that hair samples of 78% of the local population consuming fish from local sources had mercury levels three times the maximum limits allowed.[3] Likewise, illegal mining produces toxic waste—especially cyanide and mercury—that contaminate the air, soil, and water. It is estimated that illegal mining has devastated more than 50,000 hectares of forests, as well as lagoons and marshes, in the central region of Madre de Dios, one of Peru's four Amazon regions. It has even occurred in 17 PNAs and five buffer zones, as listed in Table 4.2:

That destruction spurred enactment of the Law on Formalization and Promotion of Small and Artisanal Mining (law 27651), which makes the GORE responsible for controlling and sanctioning those who qualify in the law as artisanal or small-scale mining producers, regardless of whether they are administratively accredited or not as such. Subsequently, Legislative Decree No. 1101 was issued to regulate the GORE's obligation to include, in its Annual Environmental Control

Table 4.2 Protected areas and buffer zones where illegal mining has been documented

Protected Natrual Areas (PNAs)	*National Park and Reserve Buffer Zones (around PNAs)*
1. Parque Nacional Huascarán 2. Santuario Nacional Tabaconas Namballe 3. Santuario Nacional de Machu Picchu 4. Reserva Nacional de Salinas y Aguada Blanca 5. Reserva Nacional de Calipuy 6. Reserva Nacional Allpahuayo Mishana 7. Reserva Paisajística Nor Yauyos – Cochas 8. Reserva Paisajística Subcuenca del Cotahuasi 9. Reserva Comunal El Sira 10. Reserva Comunal Amarakaeri 11. Ex Zona Reservada Gueppi 12. Zona Reservada Santiago Comaina	1. Parque Nacional Cerros de Amotape 2. Parque Nacional del río Abiseo 3. Parque Nacional Bahuaja Sonene 4. Parque Nacional Cordillera Azul 5. Reserva Nacional Tambopata

Source Created by the authors

Plan (PLANEFA for its Spanish acronym), oversight actions regarding small-scale and artisanal mining activities under its jurisdiction. This standard also requires the GORE to present quarterly reports to OEFA on their activities under the PLANEFA, as well as of non-programmed activities. Failure to submit the report is a breach that OEFA reports every 6 months to the General Comptroller.

OEFA has continuously overseen the Amazon Region GORE on their control of mining. On that activity, in 2013 OEFA conducted 11 actions in the region, and 18 each in 2014 and 2015. For its 2015 report on the 25 GORE (Organismo de Evaluación y Fiscalización Ambiental (OEFA) 2016c), OEFA examined the EIEs' four obligations on mining: (i) identification of illegal mining activities within their jurisdictions, (ii) communication of illegal mining activities to the Public Ministry and/or the Peruvian National Police (PNP), (iii) identification of cases of illegal mining and information to the GORE's Public Prosecutor, and (iv) request for legal interdiction through the Public Prosecutor's Office. It found that of the nine Amazon EIEs, seven identified illegal mining activities within their jurisdictions and five reported cases of illegal mining to the Public Ministry and/or PNP, but only one reported cases of illegal mining to the GORE's Public Prosecutor and only one requested

Table 4.3 EIE fulfillment of legal steps in environmental enforcement

Step	*Indicator*	*EIEs fulfilling their duties*	*Number*	*% Fulfilling*
1	Identification of illegal mining activities within their jurisdictions	Amazonas, Cusco, Huánuco, Madre de Dios, Pasco, San Martín, Ucayali	7	77.7
2	Identification of illegal mining activities and information to the Public Ministry and/or the Peruvian National Police (PNP)	Amazonas, Cusco, Pasco, Ucayali, San Martín	5	55.5
3	Identification of cases of illegal mining and information to the GORE's Public Prosecutor	San Martín	1	11.1
4	Request for interdiction actions through the Public Prosecutor's Office	San Martín	1	11.1

Source OEFA

legal interdiction through the GORE's Public Prosecutor. Clearly, the process of enforcement fell quite short, as illustrated in Table 4.3:

In addition to a lack of legal action, the biggest weakness was control of small-scale and artisanal mining, which is a region-level responsibility. Many reasons explain this weakness, such as the limited funding, logistical support, technical equipment, and of well-trained professionals who were, to put it generously, often less than fully aware of their responsibilities. OEFA tries to plug these holes. In 2015, for example, it provided 20 on-site visits of technical assistance as well as regional workshops in Iquitos, Cusco, and Junín.

A second area of OEFA sight is forests. Peru is heavily forested: humid tropical forests cover 56.09% of its territory,[4] providing a large number of products, especially wood. All activities affecting these forests are regulated by: (i) the Forest and Wildlife Law (Law 29763), which created the National Forest and Wildlife Management System (Sinafor) to integrate policy, norms, management tools, and relations between State institutions, the private sector, and civil society, over forest management and wildlife. This law also established a Governing Body for this system, the National Forest Service and Wildlife (Serfor); (ii) Rules for the Management of Forest Plantations and Agroforestry

Table 4.4 Production of palm oil: Colombia, Ecuador, and Peru (2012)

Country	*Production of palm oil (in tons)*	*Total cultivation area (in hectares)*	*Cultivation area within the Amazon (in hectares)*
Colombia	966,900	452,435	n/d
Ecuador	325,000	280,000	25,000
Perú	130,000	60,000	60,000

Systems, and (iii) Rules for Forest and Wildlife Management in peasant communities and native communities. These regulations establish two ways to legally access forest resources. The first are through enabling certificates, which allow access through management plans for the sustainable use of forest resources and ecosystem services, such as concessions and permits in private lands. The other are other administrative acts authorizing deforestation and changes of land use for agricultural purposes.

Despite this regulatory framework, Peru's government has not yet formulated a comprehensive solution to the spread of illegal logging, defined as any activity using commercial timber or to eliminate the vegetal cover, for commercial or non-commercial purposes, without permits, concessions, forest authorizations, and/or approved management plans (Bernales Alvarado 2008, p. 4). As described in this book's other chapters, there are many sources of illegal logging, ranging from structural problems like poverty and crime to poor implementation of environmental regulations. In the Amazon, it has proliferated primarily in tandem with oil palm cultivation, which is expanding globally to meet increasing demand for palm oil for food processing, cosmetics, and biodiesel. Great tracts of land in the Amazonia of Colombia, Ecuador, and Peru are now allocated to oil palm cultivation, as seen in Table 4.4:

In Peru, palm oil cultivation has become a major agro-industrial activity that must comply with two primary legal requirements: suitably qualified agricultural land, depending on the classification of land use (otherwise, an authorization for changing in land use must be obtained); and environmental certification of the project's specific activities. The EIEs responsible for overseeing oil palm cultivation and all agro-industry in the Amazon are Regional Governments, which approve and monitor changes in land use; and the Ministry of Agriculture and Irrigation (Minagri) to oversee environmental certification.

OEFA verifies environmental control on illegal logging by all other agencies. For the GORE, OEFA conducted eight supervisions in 2014 and two in 2015. As with mining, these actions revealed that the GORE's performance on forestry has also been woefully inadequate. Such inadequacy was exemplified by one of Peru's most notorious cases of deforestation, which took place in the Tamshiyacu Cacao Country State in the Province of Maynas of Loreto Department. After acquiring the Tamshiyacu Country State, a company called Cacao del Perú Norte S.A.C. slashed the area's forest cover in order to carry out agro-forestry cacao cultivation. But it never requested environmental certification from Minagri or authorization to change the land use by Loreto's GORE. After being alerted of this violation, OEFA repeatedly requested information from both Minagri and the GORE about their obligations. In November 2014, while providing technical assistance to Minagri, OEFA discovered that the company was operating without the required environmental management, and so ordered the company to halt activities. Subsequently, OEFA requested information on the controls developed by Minagri in order to verify the compliance with the preventive measure imposed, as well as on environmental controls on other companies related to Cacao del Perú Norte S.A.C. In March 2015, Minagri conducted a special oversight action on the Tamshiyacu Country State, with OEFA and the GORE of Loreto. In this action, OEFA found that Cacao del Perú Norte S.A.C. had not complied with the preventive measure ordered by Minagri and, in addition, was preparing new deforestation actions without authorization. OEFA then requested that Minagri carefully oversee the preventive measure and better communicate to the GORE of Loreto and Serfor about their legal responsibilities. Since then, OEFA continues to followup on the case through oversight actions to Loreto's GORE. Based on this experience, OEFA has intensified its oversight of EIEs, and, as in the mining sector, is stepping up its technical assistance to these entities.

A third area of OEFA oversight is solid waste, defined as substances, products or by-products in solid or semi-solid state that are disposed according to national regulations or their levels of health and environment risk. Solid waste is classified into four types, according to origin. The first is Household waste, generated in domestic activities; the second is Commercial waste from businesses providing goods and services; and the third is Waste generated in construction and demolition works. The fourth, which has proven to be a growing problem throughout the region, is Hospital waste generated by medical care and research in hospitals, clinics, health centers, and laboratories, among others. Such wasted is contaminated with infectious agents or may contain high concentrations of dangerous microorganisms. To control the threat to public health from such waste, the State has developed structures and laws to limit their inefficient management and disposal.

Several EIEs are involved in the management of solid waste, including Provincial Municipalities; Regional Governments (GORE); and the Health Ministry's General Body of Environmental Health (Digesa). Most of this responsibility falls on the GORE, which the General Law of Solid Waste (Law 27314) requires to monitor the proper management of solid waste from municipal sources and health facilities at all stages—collection, transportation, and especially disposal (Organismo de Evaluación y Fiscalización Ambiental (OEFA) 2016a, p. 11). The provincial municipalities, responsible for household and commercial waste, must plan, promote, regulate, approve, authorize, supervise, and sanction aspects regarding municipal waste—above all, by making sure that it gets to landfills. For its part, Digesa approves environmental studies and provides technical opinions for municipal infrastructure projects. Following up on its responsibility to approve environmental certification, Digesa also supervises the implementation of every EIS. In its oversight of the EIEs, in 2015 OEFA made 31 routine supervisions in the Provincial Municipalities of San Martín, Ucayali, Amazonas, Madre de Dios, and Loreto regions and a special supervision in the Provincial Municipality of Maynas. In addition, 226 training actions were carried out that year with provincial and district municipal governments in these provinces. A report on these actions (Organismo de Evaluación y Fiscalización Ambiental (OEFA) 2016b) details fulfillment of 15 environmental components comprising the obligations by provincial municipalities. As Table 4.5 illustrates, all the Amazon regional EIEs have had dismal records on solid waste.

One reason for this failure is beyond their control: nationwide, there are only 12 landfills for a population of over 33 million—an alarming infrastructure deficit. Likewise, 95% of supervised EIEs do not have a treatment plant for organic and inorganic solid waste. The resulting use of alternatives then multiplies waste's environmental harm. Solid waste disposal in open pits, litter incineration, lack of control of leachate, and

Table 4.5 Compliance ranking in the management of solid waste

EIE	*Score (up to 20): 2014*	*Score (up to 20): 2015*
Huánuco	3.50	8.00
Loreto (Maynas)	9.00	8.00
Cusco	7.00	7.00
Amazonas (Chachapoyas)	7.50	6.00
Pasco	7.00	6.00
Junín (Pasco)	7.00	6.00
Madre de Dios (Tambopata)	8.00	5.00
San Martín (Moyobamba)	8.00	5.00
Ucayali (Coronel Portillo)	5.00	3.00

Source OEFA

dumping of solid waste in bodies of water all adversely affect the quality of water, air soil, and Peruvians' health. For these reasons, OEFA helps local governments to prioritize the treatment, reuse, and recycling of solid waste; to promote segregation through the differentiated collection of organic and inorganic municipal solid waste; and to adopt technologies that generate energy from solid waste.

Illustrating the challenges involved in such promotion was the "El Treinta" landfill in the province of Maynas, department of Loreto, within the buffer zone of the Protected Natural Area Allpahuayo-Mishana. This landfill was run by the company BRUNNER S.A.C., which was inadequately maintaining solid waste and thus risking the surrounding population's health. Elevating those risks was the EIE's failure to detect these actions, which turned out to be one link in a chain of failure: the Provincial Municipality of Maynas did not comply with monitoring the municipal solid waste management; Digesa did not verify compliance with the obligations stipulated in the approved EIS; and, in turn, the Regional Government of Loreto did not monitor solid waste destined for the landfill. OEFA sent a report to the Peruvian General Comptroller to explain the EIE's failure and help determine applicable penalties.

Environmental Control of Companies

In Peru's Amazonía, many companies work in the sectors that OEFA oversees. Among them, the hydrocarbons sector has the largest number of companies operating in the Amazon region (31 companies), followed by the electric sector with 24 companies, the mining sector with 13, the fishing sector with five, and the industry sector with three. Of all those businesses, those with the greatest environmental impact in the area are those engaged in the exploration, extraction, and transportation of hydrocarbons. In response, OEFA has sharpened its monitoring of these activities, which led it to be directly involved in several major cases.

Case 1: Oil spill in the KM 41 + 833 ONP: On June 30, 2014, Petróleos del Perú S.A.—Petroperú (PETROPERU) reported an environmental emergency from an oil spill at km 39 + 584 of the Norperuano Oil Pipeline. In response, OEFA officials immediately went to the area to verify the magnitude of the impacts. In their first three supervisions, they found a huge amount of dead fish and destroyed vegetation in the vicinity of the pipeline breach. A report was then issued detailing violations related to the spill. Subsequently, Petróleos del Perú S.A—Petroperú was found responsible, and a corrective measure was ordered to restore the impacted area back to its natural state.

Case 2: Preventive measure for soil removal: There are also cases in which OEFA acts without such an emergency, such as with preventive measures to avoid an imminent danger or high risk of serious harm to the environment, natural resources, and human health, as well as to mitigate the causes of damage that does occur. An example of such measures was during an environmental supervision carried out from March 2 to 7, 2013, in which OEFA documented soil infused with hydrocarbons in the area called "Los Jardines" in the influence area of the Block 1-AB, owned by the Pluspetrol Norte S.A. company. In its soil removal activity, the company was mixing impacted soils with native soils. Such mixing was not approved by any agency, and probably would not be, since it risked spreading contaminated soil into areas larger than those initially impacted. In order to avoid such damage, a Directorial Act that ordered an immediate cessation of soil removal work in "Los Jardines".

Two additional cases of control over mining and hydrocarbon activities in the Amazon further illustrate the complexity of environmental enforcement in the region, particularly in punishing abuses. One case involves the many forms of illegal mining, which OEFA works to control. Law No. 29325 states that when OEFA obtains reasonable and verifiable proof of non-compliance with the conditions for an activity within the purview of regional governments, it has the power to carry out environmental controls. In utilizing this power, in 2014 OEFA approved the Legal Rules for the application of Article 17 of Law 29325 to allow it to determine the extent of questionable mining activities and those conducting them. Pursuant to these rules, OEFA may initiate a sanctioning procedure when it obtains reasonable proof that someone is pretending to be a small-scale miner or artisanal miner without certification. Up until 2015, OEFA dictated 23 sanctioning procedures in illegal mining, of which five were in the Amazon region (Madre de Dios, Cusco, Puno, and Huánuco). An example was OEFA's sanction on Yanguang Yi in Puerto Inca, Huánuco, for illegal mining activities. A fine of 498.19 UIT (more than US$500,000) on Yi was accompanied by a cessation order on activities until full compliance with legal requirements.

Because of the conflict it sparked, a more well-known case was in the basins of the Pastaza, Corrientes, Tigre, and Marañón rivers, in the department of Loreto. In that area are two of the country's biggest oil fields, Block 192 (formerly Block 1-AB) and Block 8. When oil deposits were discovered in the 1970s, Block 192 was given to Occidental Petroleum Corporation (Oxy) and Block 8 to Petróleos of Peru S.A—Petroperú (PETROPERU). In 2001, Pluspetrol Peru Corporation S.A. entered the area to operate Block 192, and 2 years later the company split into two, leading to formation of Pluspetrol Norte S.A., called Pluspetrol, which was subrogated in the position of transferee in the concession contract, with a clear expiration date of August 29, 2015, for the exploitation of Block 192. The operation of Block 192 by Pluspetrol caused constant environmental complaints by local populations, who began to demand remediation from over 20 years of damage from these oil operations. In 2006, a Supreme Decree established the Complementary Environmental Plan (PAC) for companies carrying out hydrocarbon activities, in order to mitigate environmental impacts unaddressed within the time and under the requirements in the companies' respective Environmental Management Plans (PAMA). The 2005 PAC for Block 1-AB, operated by Pluspetrol, required repair in areas of operation not considered in the 1996 PAMA, which would have to pass government approval, and made the company liable for damages. Through it, Pluspetrol was required to identify previously impacted sites and, subsequently, to remedy the 75 identified sites impacted with hydrocarbons. This PAC was in force until 2009, by when Pluspetrol had to comply with remediation of the 75 identified sites. But because the company did not comply in nine of those sites in time, in 2015 it was forced to submit a Cessation Plan for non-compliance.

Due to the demands of the populations affected by contamination of the four basins, meanwhile, the Multisector Commission attached to the Presidency of the Council of Ministers (Multisector Commission) was formed in 2012 to analyze and propose measures to address the adverse impacts on the area's residents. Through the Multisector Commission, MINAM declared the Pastaza, Tigre, and Corrientes river basins—in particular, areas located within Block 192—to be in a state of environmental emergency. This Commission had two working groups: the Environmental Working Group and the Social Working Group. OEFA was part of the Environmental Working Group implementing the Action Plan for each of the Environmental Emergency Declarations (DEA for its Spanish acronym). In that role, OEFA conducted investigations in the area, identifying 92 impacted sites that had not been included in the Block 192 PAC. A report on the environmental situation of Block 1-AB, operated by Pluspetrol, included results of actions carried out as well as analysis of impacted sites not included in the PAC. OEFA directed the company to immediately remediate these areas through an IGA known as "Abandonment Plan," a more comprehensive remediation instrument that requires abandoning operations and the concession area to allow for cleaning and environmental remediation. On January 29, 2015, Pluspetrol submitted an Abandonment Plan, but it was rejected for not considering all environmental impacts. In February 2015, the Investigation Office of OEFA's Directorate of Sanctions initiated a sanctioning procedure against Pluspetrol for breaches to environmental regulations and to its own management plan. In 2015, as part of this procedure, the Directorate ordered Pluspetrol to identify all areas impacted by hydrocarbons and to include them in a remediation plan in accordance with environmental regulations. These actions were to be implemented unless the certification authority indicates otherwise in the procedure for the approval of the Abandonment Plan of Block 1-AB.

Conclusions and Challenges for Environmental Enforcement in the Amazon

As detailed above, OEFA is constantly supervising national, regional, and local EIE in the Amazon, in order to verify fulfillment of their environmental control responsibilities. Seeing chronic deficiencies in those responsibilities, OEFA is increasing its technical assistance. After all, it seeks to avoid being purely coercive, focusing less on reprimanding the EIEs and more on guiding them to perform better over time with cooperative and innovative practices. But the goal of enforcement of standards in the Amazon, by OEFA and others, is a constantly shifting and growing target. One particularly difficult challenge is illegal mining, especially the control of small-scale and artisanal mining overseen by the GORE. In several oversight actions, OEFA noted that most of these EIEs do not initiate sanctioning procedures when noticing failure to comply with environmental obligations. Proposing that OEFA assume environmental control of the small-scale and artisanal mining, the Executive presented Bill No. 3478/2013-PE to strengthen administrative control over facilities that carry out small-scale and artisanal mining activities. It is expected that this Bill will be approved to improve control of the mining sector and thus prevent the informal and illegal activity damaging the Amazon.

Because sustainable development of the Amazon is one of the long-term objectives of Peru's Constitution and National Environmental Policy, a special environmental protection regime has been established for the Basin to promote sustainable use of its resources. With the wide array of those resources, activities affecting each one have their own legal framework and specific authorities. But they are all backed up by a strong environmental institutionalism in the region. Overseeing OEFA and the EIEs, Sinefa's role is thus critical, because it oversees all economic activities impacting the environment. Within this structure, OEFA's work is particularly vital in the control of large extractive activities in the region. OEFA develops the full environmental control process, which includes assessing the impacts of business operations, supervising activities in order to ensure compliance with their environmental obligations, and imposing sanctions in cases of non-compliance. By encouraging voluntary rectification of environmental breaches before reaching the sanctioning procedure, OEFA also promotes timely remediation. It also issues corrective measures as a means to achieve an effective restoration of environmental damage. With these and other functions, such as providing technical assistance to the EIEs, OEFA seeks to strengthen Sinefa—and the country's wider environmental guarantees—through harmonious, coordinated, and effective environmental control by every state agency, over every economic activity, and in every corner of Peru's Amazon.

Notes

1. The Convention on Biological Diversity (CBD) defines the protected area as a geographically defined area that is designated or regulated and managed to achieve specific conservation objectives.
2. Reserved areas are those that meet the conditions to be considered PNA and require a study to determine their extent and categorization.
3. Study available at: http://www.minam.gob.pe/mineriailegal/los-efectos-de-la-mineria-ilegal/.
4. The forests of the "low jungle" cover 43.6%, "high jungle" forests 9.46%, Andean forests 0.17%, and coastal forests 2.86% (Ministry of Environment 2012, p. 98).

Bibliography

Bernales Alvarado, Manuel Ernesto. 2008. *Informe sobre la Tala Ilegal en el Perú. Visión y Aportes de la Presidencia de la Comisión Multisectorial de Lucha contra la Tala Ilegal.* Lima: Consejo Nacional del Ambiente.

Gómez Apac, Hugo R., and Milagros Granados Mandujano. 2015. El macroproceso de la fiscalización ambiental. *Revista de Derecho Administrativo: Derecho Ambiental* 15: 75–95.

Juan, Dammert B. 2014. *Cambio de uso de suelos por agricultura a gran escala en la Amazonía Andina: el caso de la palma aceitera.* Lima: SPDA, USAID, and Iniciativa para la Conservación en la Amazonía Andina.

Lozano, Blanca. 2010. *Derecho ambiental administrativo.* España: La Ley.

Ministry of Environment. 2012. *Mapa Nacional de Cobertura Vegetal. Memoria Descriptiva,* 2012. Lima: OEFA.

Ministry of Environment. 2016. *La fiscalización ambiental en el Perú (2011–2015).* Fortaleciendo los cimientos del derecho a un ambiente sano. Lima: Ministry of the Environment, Republic of Peru.

Organismo de Evaluación y Fiscalización Ambiental (OEFA). 2015. *La vinculación y retroalimentación entre la certificación y la fiscalización ambiental.* Lima: OEFA.

Organismo de Evaluación y Fiscalización Ambiental (OEFA). 2016a. *Fiscalización Ambiental del Sector Salud a nivel de gobiernos regionales. Informe 2015. Índice de cumplimiento de los gobiernos regionales.* Lima: OEFA.

Organismo de Evaluación y Fiscalización Ambiental (OEFA). 2016b. *Environmental control to small-scale mining and artisanal mining, report 2015: Compliance index of regional governments.* Lima: OEFA.

Organismo de Evaluación y Fiscalización Ambiental (OEFA). 2016c. *Environmental inspection in solid waste to provincial municipalities, report*

2014–2015: Compliance Index of provincial municipalities at the national level. Lima: OEFA.

Programa de las Naciones Unidas para el Medio Ambiente (PNUMA), Organización del Tratado de Cooperación Amazónica (OTCA) and Centro de Investigación de la Universidad del Pacífico (CIUP). 2009. *Perspectivas del Medio Ambiente en la Amazonía: Geo Amazonía.* Lima: PNUMA.

Sociedad Peruana de Derecho Ambiental (SPDA). 2014. *La realidad de la minería ilegal en países amazónicos.* Lima: SPDA.

CHAPTER 5

Ecuador: Rainforest Under Siege

Víctor López Acevedo

Abstract This chapter explains the tightening pressures on the many dimensions of protection in the Ecuadorian Amazon, from conservation of its biodiversity to repelling multiple incursions of its elaborate system of protected areas, the anchor of the country's vaunted constitutional protection of Nature's rights. With a focus on the extensive oil concessions and incursions on protected areas, it critiques the disparity between the country's strong ecological movement, infrastructure, and law on the one hand, and, on the other, the economic priorities of a government intent on exploiting the Amazon's resources.

Keywords Amazon · Environment · Indigenous peoples · Nature reserves

Translated by Mark Ungar.

V. López Acevedo (✉)
Corporación Pro-Ideas, Quito, Ecuador

M. Ungar (ed.), *The 21st Century Fight for the Amazon*,
https://doi.org/10.1007/978-3-319-56552-1_5

Introduction: What Is Tropical Deforestation and How Can It Be Measured?

This chapter examines the multiple strains on the Ecuadorian Amazon Region, focusing on the extractive economy concessions as drivers for forest lost and their incursions on protected areas and indigenous territories. Through four key questions, it assesses the disparity between the country's strong ecological movement, infrastructure and law—rooted in the world's first constitutional protection of nature—and the economic development priorities of a government with no consideration for political opposition claims but strong incentive to exploit the Amazon basin's resources.

Measurements of deforestation are continually being conducted in each region of the Amazon basin—the Western or continental Amazonia, as it is known in Brazil; the areas of geological transition between the Andes and the non-flooding Amazonia; and the Guyanese sub-region where the Orinoco and Amazon river basins overlap (RAISG 2012, 2015). In each of these sub-regions, deforestation is reported as areas where forest was lost in determined periods of time (2000–2005, 2010–2015, etc.), and which can be cataloged using remote sensing as "recent", clear-cut areas where the ground is still exposed, or "prior," where natural dynamics show signs of regeneration with herbaceous or pioneer vegetation that have not yet reached a secondary growth forest cover, as in much of the Ecuadorian Amazon Region (*RAE: Región Amazónica Ecuatoriana*). As with any other sub-region of the Amazonas-Orinoco macro-basin, the RAE's tropical deforestation is the result of a long process of land use change driven by mercantile and institutional factors, from export demands to rural poverty, which together speed up loss and degradation of the forests as well as the vulnerability of the peoples who depend on them. The wider definition of "forests" used in this chapter adopts the criteria established by international organizations and instruments (such as the FAO and UNDP), that define them as "land that cover more than 0.5 ha, with trees higher than five meters and with a coverage of canopy of ten percent, or trees capable of reaching these minimum limits in situ."

One of the advantages of generating and analyzing comparable data for all the countries of the Amazon macro-basin is that it allows a comparative approximation of the magnitude of the phenomenon in equivalent units of analysis. In that way, RAISG and its Deforestation Working

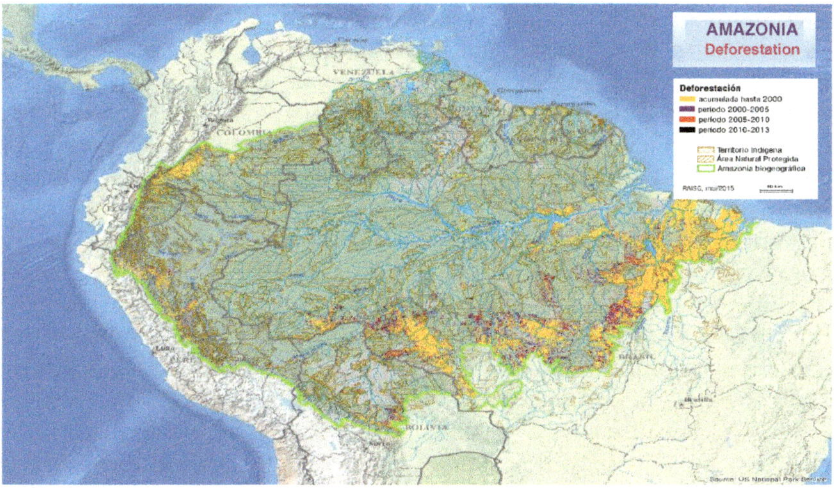

Fig. 5.1 Deforestation in the Amazon (1970–2013)

Group needed to first validate preliminary information, mostly for the main units of analysis of this study: Natural Protected Areas (ANP: Áreas Naturales Protegidas), Indigenous Territories (TI: Territorios Indíginas), and hydrographic units or river basins. This allowed to understand patterns of tropical deforestation in the macro-basin as a whole as well as in each of the countries that comprise it (RAISG 2015), as illustrated in the pan-Amazonia Fig. 5.1.

Deforestation in the Amazon's forests was highest prior to 2000, and although it subsequently declined, affected surfaces continue to be critical in different countries (see Fig. 5.1), mainly due to national policies that promote the exploitation of non-renewable natural resources, particularly primary and energy materials.

This destructive process is most apparent in the "Arc of Deforestation" carved out by the growth in the economy of scale of agro-industrial uses heavily promoted in extensive zones of Brazil, Bolivia, and Peru. This expansion is on top of long-standing forms of "prior deforestation" that persist in the Andean-Amazon Piedmont, primarily in the forms of subsistence agriculture promoted by agrarian and colonization legislation—such as enacted in Ecuador in 1964, 1973, and 1994. Another increasing cause of destruction, as increasingly documented by the United Nations and governments, is illicit narcotics cultivation and transit connected to the processing of drugs. Such cultivation is expanding in many areas of the Amazon, such as border zones between Colombia and Ecuador, in Peru's central jungle and from the tropical valleys known as *yungas* through to the Chapare region of Bolivia (FAN 2015). North of the macro-basin, deforestation is associated with the expansion of aluminum and gold extraction, as well as the opening of roads required, and which in turn opens up yet additional fronts for logging, scarring TIs in Guyanas, the south of Venezuela (RAISG, *op. cit.,* 2012) and to a lesser extent, the country of Surinam. In the Ecuadorian Amazon, deforestation in recent years was linked to the appreciation in value of primary materials during the decade from 2005 to 2014, which fueled the expansion of mineral and hydrocarbon extraction, but also of wood, palm oil agro-industry, while maintaining inefficient systems of cattle ranching and livestock farming, above all beef production.

What Are the Direct and Indirect Drivers of Deforestation?

Among direct causes of deforestation are a series of state policies aiming to meet the international demand for commodities and energy materials and flows, fueled by a flurry of subsidies and speculation. Production of commodities or raw materials with no added value (e.g., soy, oil palm, meats and other primary products) is another driver of deforestation in the macro-basin. The governments of every Amazon country bend over backwards to meet the demands of the international market—even to the detriment of environmental policy and indigenous territorial rights, and even amid price and market volatility (as seen in the rapid decrease of demand from China) that leads these exports to be labeled "risk commodities". Demand for energy, minerals and other primary materials such as the Amazon crude of Ecuador and Colombia; aluminum and iron from the south of Venezuela, Guyana and the north of Brazil; as well as gold from the Amazon rivers of Peru, are another cause of tropical deforestation, all associated with the extraction of subsoil resources, which is the exclusive domain of the states but in practice with little actual state control. Another large international market, of course, is the illicit one. Throughout the Andean-Amazon sub-region, deforestation has been also spread through coca cultivation for international narcotrafficking, as well as by fronts of illegal mining and chains of illegal timber extraction, especially in the frontier zones where the Colombian department (province) of Putumayo borders Ecuador, as well as in the central jungles in Peru and in Bolivia's Chapare region. These activities expose the inherent weakness of national institutional control, particularly of local-level governance.

As in other countries, deforestation in Ecuador has also been spurred by national plans to use Amazonian space for the transfer of landless people from other regions. In fact, a legal requirement for the titling of land is to strip down half of the allocated property, in the classic formula known as "land for deforestation". As in other countries as well, such transfer was used to shore up control of the remote frontier zones, particularly after the 1941 and 1995 armed conflicts with Peru, triggered in part by Ecuador's long-standing claim to a huge chunk of Peruvian Amazonia. This range of policies and institutional expansion for the colonization, militarization, and agro-industrial promotion of the Amazon then drives an increasingly rapid pace of deforestation. In the Amazonian regions of Ecuador and other countries, that process usually starts with selective extraction and logging that transitions to open-soil operations. A slash and burn preparation is followed by short-term cultivation for subsistence, followed by perennial crops for the world market, which in turn often lead to development of pastures for unproductive livestock ranching involving soil depletion, poor management, and other climate determinants (e.g., humidity) and topographical accessibility. This transformation also spurs the paving and opening of major routes, which is one of the principal drivers of tropical deforestation not only by facilitating new extractive fronts (mining, timber or hydrocarbons), but also for the establishment of lines of colonization through the allocation of lands and the Basin-wide "fishbone" pattern of unauthorized secondary roads.

In order to attain energy security, several hydroelectric projects have been established in the Andean-Amazon piedmont of Ecuador, and above all in the flooding plains of Peru, Brazil, and Bolivia. Carved up to allow the passage of high-power transmission lines, these flood-prone areas will be affected by the loss of natural vegetal cover in what has been deemed as expansion of "non-consumptive frontiers" in Amazonia (López et al. 2016). In addition, through various initiatives for integration of transport infrastructure (roads and waterways), they will connect areas of natural resource exploitation in pan-Amazonia with international markets of the Pacific and Atlantic, which demands low-cost energy. Such demand is of course another of the many catalysts of tropical deforestation, often associated with strategic projects financed as part of regional integration plans by Pan-American agencies such as Union of South American Nations (USAN), *Banco Sur,* and other binational funds.

Altogether, this range of economic policies and market processes—from international demands and energy policies to demographic politics and land settlement—reveal the steep challenges facing national institutions created for deforestation control, administration of protected areas and conservation of communal lands. The depletion of conventional hydrocarbons, as well as the entry into the international market of "non-conventional" processes—such as fracking from hidraulically fractured wells—entail a transformation of the economic macro-structure in countries based on exports of primary materials.

How Do We Best Understand Tropical Deforestation in the RAE?

RAE has been defined as a sub-national political-administrative division with a surface of 116,588 km^2, which is 47% of national territory. If the hydrographic slope to the Amazonas in Ecuador is included, that total area expands to 131,839 km^2, equivalent to 52% of the country (López et al. 2013). However, if only ecosystems that correspond to vegetable formations or strictly Amazon forests are included, these areas are reduced to an estimated 103.226 km^2, an area that in 2000 was then reduced to 87,053 km^2 of remaining original forest, as was reported by recent RAISG studies (RAISG 2015; López et al. 2015 and actualization analysis).

Although the RAE has less than 6% of the national population, among the nine Amazon countries, the Ecuadorian Amazonia is the most densely populated, with an estimated 6.4 habitants per square kilometer (López et al. 2013). As of 2010, the RAE population that recognizes itself as indigenous comprised about 250,000 persons, distributed in 11 different nationalities: Kichwa, Shuar, Achuar, Waorani, Sapara, Siona, Secoya, Cofán, Shiwiar, Andoa and Quijos. Also concentrated in the RAE are special areas for forest protection and other

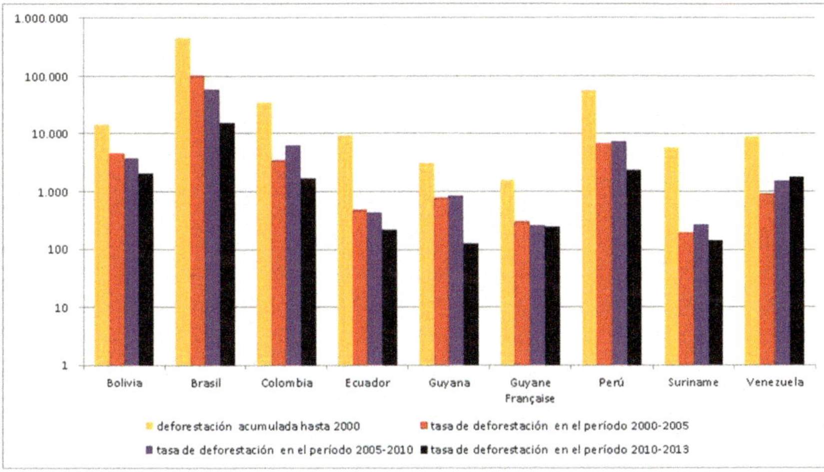

Fig. 5.2 Historical and recent deforestation in the Amazon. *Source* RAISG (2015)

aquatic ecosystems of the Amazonia, such as the Protected Natural Areas (ANP: *Areas Naturales Protegidas*), which includes units of the National System of Protected Areas (*SNAP: Sistema Nacional de Áreas Protegidas*) with 31,331 km^2 and Forests and Protective Vegetation (*BVP: Bosques y Vegetación Protectores*) with an additional 9.238 km^2, bringing the total area up to 40.569 km^2 (López et al. 2016). Indigenous territories (*TI*) must also be included, with an extent calculated in 2014 at 65,243 km^2, which is more than 56% of the entire Amazon region. Of the TI land, an estimated 19,212 km^2 overlaps (double and multiple) with the conservation zones, as reported in an updated study (see Fig. 5.2).

The Amazon forests within and above all out of these special areas, however, are subject to growing pressures and threats by a series of socio-environmental patterns. Forest loss, whether recent or prior deforestation, as well as the degradation or fragmentation of ecosystems, are the corollary of an extractivist economic approach (more than productive or industrialized), characterized by a type of economic income that is rent-based and degradative, with high environmental and human impact in the areas of natural resource exploitation.

With its location in the Andean–Amazon sub-region of the macro-basin, the RAE is comprised of an area of transition between the Andes and the Eastern Mountain Range (*Cordillera Real Oriental, CRO*), defining piedmont ecosystems or mountain slope (up to 2.300 meters above sea level), upper Amazon jungle (between 2.299 and 600 meters above sea level), as well as a small part of the flooded Amazon plains or low jungle (<600 meters above sea level). It is still pending the calculation of areas and rates of deforestation in those sub-regions, which is difficult because analyses must be adjusted to each one of these vegetation formations and forests.

To protect its Amazon region, and to lay a foundation for its breakthrough constitutional protection of Nature, Ecuador created a National System of Protection Areas (SNAP: *Sistema Nacional de Areas Protegidas*). SNAP is comprised of four sub-systems, each with specific regulations: the Federal State Patrimonial Subsystem of Natural Areas (PANE: *Patrimonio de Áreas Natural del Estado*), comprised of 48 protected areas covering approximately 20% of national territory, one of Latin America's highest rates; the APG (*Áreas Protegidas de los Gobiernos Autónomos Descentralizados*), run by the country's 24 provinces; privately owned Private Protection areas (APPRI: *Subsistema de Áreas protegidas privadas*); and communitarian Protected Areas (APC: *Áreas Protegidas Comunitarias*), which covers indigenous (TI: *Tierras Indígenas*) and Afro-Ecuadorian lands.

Table 5.1 presents aggregated results for areas and rates of deforestation in the RAE, based on the SNAP system; note that in all the period and units of analysis (ANP, TI, and hydrographic basins), deforestation is wider in areas that have no type of protection (López et al. 2016): The above-mentioned case study identifies that, in 2013, Ecuador's remaining Amazon forests would cover an estimated 87,060 km^2

ANP and TI: Barriers to Deforestation in the Pan-Amazonia…and Ecuador?

Because of the accelerating changes in land use and the loss of forest cover, protected areas (*ANP*) and indigenous territories (*TI*) have become "forest islands" in many areas of Pan-Amazonia, subject to constant external pressures in a territorial mosaic characterized by the fragmentation of ecosystems, the degradation of forest and environmental deterioration by unsustainable production, or primary extraction of non-renewable resources. Territorial security of land allocation in favor

Table 5.1 Deforestation in the ecuadorian Amazon region (RAE)

			Deforestation rate			*% of the original forest*		*% of remaining forest from 2000*
	Surface of original forest cover	*Cumulative deforestation until 2000*	*2000–2005*	*2005–2010*	*2010–2015*	*2000–2015*	*Cumulative total*	*2000–2015*
	km^2	*km^2*	*km^2*	*km^2*	*km^2*	*%*	*%*	*%*
Ecuadorian Amazonia (biogeographic)	**96,073**	**9020**	**1054**	**1090**	**957**	3.2	12.6	3.5
Outside of *ANP* and *TI*	**20,880**	**5911**	**476**	**540**	**457**	7.1	35.4	9.8
Indigenous Territories (*TI*)	**61,209**	**2816**	**482**	**427**	**356**	2.1	6.7	2.2
Occupation Traditional without recognition	1164	11	5	15	16	3.2	4.1	3.2
TI officially recognized	58,738	2805	457	410	339	2.1	6.8	2.1
Intangible Zones (*ZI*)	1262	0	20	2	1	1.8	1.9	1.9
Protected Natural Areas (*ANP*)	**29,766**	**480**	**240**	**206**	**199**	2.2	3.8	2.2
National-indirect use	29,640	472	239	205	197	2,2	3,8	2.2
Departamental (*GAD*)-indirect use	126	7	1	1	2	3.0	8.7	3.2

Source Based on data from RAISG (2015) and López et al. (2015) and actualization analysis

of indigenous people, as well as *ANP* declaration that make up national protection and conservation systems, are also under pressure and threat by extractive fronts, beyond titling or delimitation of the respective areas. And when legal boundaries of indigenous land have not been settled as communal property, or when units of conservation are not clearly demarcated, their vulnerability is greater, resulting from its economic status as "common pool resources" of open availability—a free-for-all without restrictions on access and extraction. In addition, public domain may affect communal property or *ANP* status when a government declares extractive projects within a *TI* or *ANP* as "national priority" due to energy security or fiscal reasons.

Population is another determining variable. On one end of the spectrum, although parks without people are an exception in countries like those in the Andean-Amazon sub-region or a good part of the continental Amazonia and the Guyanese Plateau, they still raise new challenges for monitoring and control. On the other end, if the overlapping areas of *ANP* and *TI* have traditional settlements (beyond property titles) of populations with their own uses of land and water ecosystems, the definition of fixed limits thus presents a challenge for territorial management. The role of population is particularly relevant in the Ecuadorian Amazon, which has Pan-Amazonia's highest population density.

The first *ANPs* in Ecuador's Amazonia were established by dictatorial and military regimes in the 1970s, at a time when the country lacked specific legislation, or even a clear policy framework in any issue. That changed with the 1981 Forest Law, which began to set out the legal instruments, still in force, for protection of natural areas, (Table 5.2), forests, and protective vegetation, as well as other new forms of forest conservation (Table 5.3):

The cited study warns of a tendency of decreasing areas of the ANPs declared from 1970 to 2014, as can be seen in Fig. 5.3. For PANE units, units comprising two-thirds of the current area of the sub-system were formed between 1975 and 1979; and if the first unit created during the previous period is added, then three-fourths of the current total ANP area was founded during the 1970s. New PANE units that followed in the subsequent five-year periods were marginal (i.e., 37 km^2 in the period of 1985–1989), with the exception of the 1990–1994 period, registering a significant contribution to the surface area of the current sub-system (3.264 km^2), as detailed in Fig. 5.4.

Table 5.2 Categories of integral porotection in the Amazonia of ecuador

Categories	*Description*	*Objectives*	*Type of use*
National park*	Extensive areas with the following characteristics or purposes: (1) One or more ecosystems with a minimum of 10,000 hectares. (2) Diversity of species of flora and fauna, geological features and habitats of importance for science, education, and recreation. (3) Maintenance of the area in its natural condition, for the preservation of ecological, aesthetic, and cultural features, prohibiting any exploitation or occupation	Protection of Wildlife, Management of Natural and Cultural Resources, Facilitation of Research, Development of alternatives to tourism	Indirect
Biological reserve*	Areas of variable extension, in any land or aquatic areas, designed for wildlife preservation and oriented to conservation of natural processes, facilitating scientific research, education, and conservation of genetic resources	Protection of Wildlife, Protection of species, Facilitation of Research	Indirect
Reserve of faunal production*	Wild flora and fauna under state dominion under MAE for it conservation, Protection and administration, for which the state exercises its policy and administrative authority	Control of the hunting, harvest, capture, transport, and trafficking of animals, amongst others established by Law and Regulations	Indirect

(continued)

Table 5.2 (continued)

Categories	*Description*	*Objectives*	*Type of use*
Ecological reserve*	An area of at least 10,000 hectares with the following characteristics or goals: One or more ecosystems with important species of wild flora and fauna, threatened with extinction, leading to prohibition of any type of exploitation or occupation. Singular geological formations in natural or partially altered areas	Protection of Wildlife, Protection of species, Natural Resources management	Indirect
Wildlife refuge*	Areas essential to guaranteethe existence of Wildlife, resident, or migratory, with scientific, educational, and recreational aims	Protection of Wildlife and other species, Facilitation of Research, Encouragement of alternatives to tourism	Indirect
Protected forests—*Bosques y Vegetación Protectores* (BVP)	Vegetal, natural, or cultivated formations with one or more of the following characteristics: (a) Have soil and Wildlife conservation as its principal function; (b) Situated in areas for water run off control or preservation of hydrographic basins, especially in zones of scarce precipitation; (c) Occupy of mountains slopes or areas contiguous to water sources, currents or deposit; (d) Constitute windbreaks or of Protection of environmental balance; (e) Found in areas of hydrological-forest research; (f) Are in strategic national defense zones; and (g) Constitute a factor of defense of Natural Resources and of infrastructure and public interest works	Activities of conservation and sustainable forest management, in accordance with management plans and their corresponding zoning of areas of use or protection, as specified in the Forest Law	Indirect/Direct

(continued)

Table 5.2 (continued)

Categories	*Description*	*Objectives*	*Type of use*
Protected areas of decentralized autonomous governments (Sub-system)	They are part of National System of Protected Areas (SNAP), but under promotion and administration of different levels of the country's autonomous decentralized government: rural parishes, cantons (municipalities), and provinces	Protection of biodiversity	Indirect
Community protected areas (Subsystem)	Form part of the National System of Areas Protected (*SNAP*) and under community promotion and administration	Protection of biodiversity	Indirect
Private protected areas (Subsystem)	Areas that are part of the National System of Protected Areas (*SNAP*), under private promotion and administration	Protection of biodiversity	Indirect

(continued)

Table 5.2 (continued)

Categories	*Description*	*Objectives*	*Type of use*
Intangible zones (*ZI*)	"Protected spaces of great cultural and biological importance in which no type of extractive activity can be carried out, due to the high value that they have for the Amazonia, Ecuador, the world and present and future generations." There are two such zones: Z.I. Cuyabeno-Imuya and Z.I. Tagaeri-Taromenane, which are integrated with two NPA: the Cuyabeno Reserve (*RPFC*) and Yasuní National Park (*PNY*), respectively. Development of policies and mechanisms of protection of the rights of indigenous peoples in voluntary isolation are permitted, such as in the case of the core of Yasuní park. They may be part of a policy of indigenous protection combined with elements of biodiversity conservation, although they are not formal parts of the *SNAP*	To safeguard the integrity of ethnic groups in living in the protected área, prohibition of petroleum and timber activities. (D.E. 551, *RPFC*) Legal guardianship of land rights for indigenous peoples in isolation, such as the Tagaeri and Taromenane groups in Yasuní park and areas of Waorani community lands (D.E. 552, *PNY*)	Indirect/strict protection for indigenous peoples in voluntary isolation (*PIAV*)

Notes *Categories that correspond to Units of Natural National Patrimony SubSystem (*PANE*). *Sources* López et al. (2016), MAE (2007), Dudley (2008), Elbers (2011), of Marchi et al. (2013), López et al. (2013), SENPLADES (2013), Designated laws (Forests, conservation of protected areas and Wildlife Act, Cod. 2004)

Table 5.3 Categories of sustainable use in Amazonia of ecuador

Categories	*Description*	*Objectives*	*Use*
National forest patrimony (*PFE*)	Areas for sustainable forest management under national authority, in this case the Agriculture and Ranching Ministry. This use has contributed to formation of a few new protected areas (ANP) in the RAE	To reserve areas for controlled forestry uses under sustainable management	Direct and Indirect
Biosphere reserves[a]	(1) Land, coastal, and marine ecosystems, or a combination, recognized in the framework of the Man and Biosphere (MAB) Program of UNESCO (2) To encourage the harmonic integration of populations and nature, in order to promote a sustainable development through participatory dialogue, interchange of knowledge, poverty reduction, welfare improvement, respect for cultural values, and the capacity of social adaptation to change	(1) To conserve landscapes, ecosystems, species, and genetic diversity (2) To attain sustainable economic and human development from sociological and ecological viewpoints (3) To support local, regional, national, and global projects of skills, education, training, research on the environment, and sustainable development	Direct

(continued)

Table 5.3 (continued)

Categories	*Description*	*Objectives*	*Use*
RAMSAR convention sites (Wetlands convention)	The Convention's mission is "the conservation and wise use of all wetlands through local and national actions and international cooperation, as a contribution towards sustainable development."	Maintain the state of conservation of the wetlands Halt or avoid activities of development the adversely affect wetlands	Direct and Indirect
Areas of the *MAE's* socio- bosque program[b]	Areas of native forests at risk of deforestation and threatened moorlands, as well as native forest that are under-represented in the *SNAP* with hydrological or biomass importance. These lands a property of high-poverty populations with high levels of poverty. The program provides an incentive for peasant and indigenous communities that make a voluntarily commitment to the conservation of their forests and moorlands	(1) To promote conservation of native forests, moorlands, and other native vegetal formations (2) To reduce GHG emissions caused by deforestation (3) To improve living standards of rural populations settled in these areas (4) To recuperate natural coverage through ecological rehabilitation through plantations of native trees or within agroforestry systems	Direct

(continued)

Table 5.3 (continued)

Categories	*Description*	*Objectives*	*Use*
Priority zones for establishment and management of connectivity corridors	Zones where corridors are needed to promote development of animal populations through coherent territorial regulation balancing development and the maintenance of natural resources. The corridors must be located in areas prioritized by the MAE as a "source" of dispersion, migration, or intermingling of species of flora and fauna that allow for the maintenance of healthy and viable populations and where the ANPs represent the "nucleus" of corridors, since they were representative specimens from the ecosystems are best conserved. Five of the 11 priority zones for establishment of corridors in Ecuador are in Amazonia	1. To establish a framework of reference to strengthen diagnostics and formulation of Development Plans and of Territorial Regulation 2. To minimize fragmentation of ecosystems, particularly of those considered fragile, endemism hotspots, groundwater recharge zones, and high genetic variability areas, which are of strategic importance to the Nation	Direct and Indirect

Sources Base don data from [a]UNESCO cited in: Ministerio del Ambiente del Ecuador. 2010. Biosphere Reserves of Ecuador: exceptional places. GTZ/GESORENDED—WCS—NCI-UNESCO/Quito. Ecuador

[b]MAE. Protecting forests and encouraging development of campesinos and indigenous peoples of the country. Programa Socio Bosque

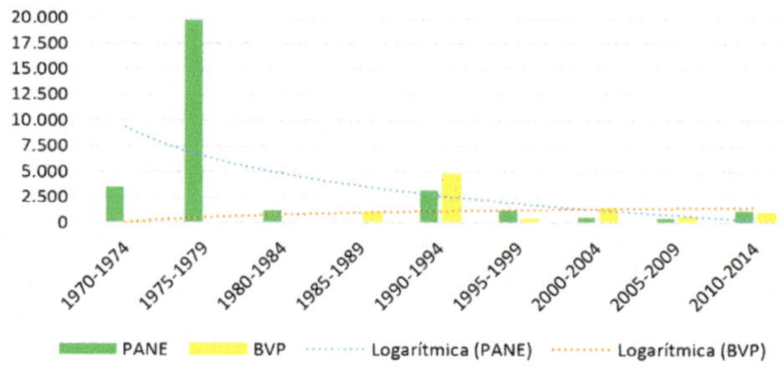

Fig. 5.3 Patterns of ANP creation, in square kilometers (km^2). (1970–2014). *Source* Created by author

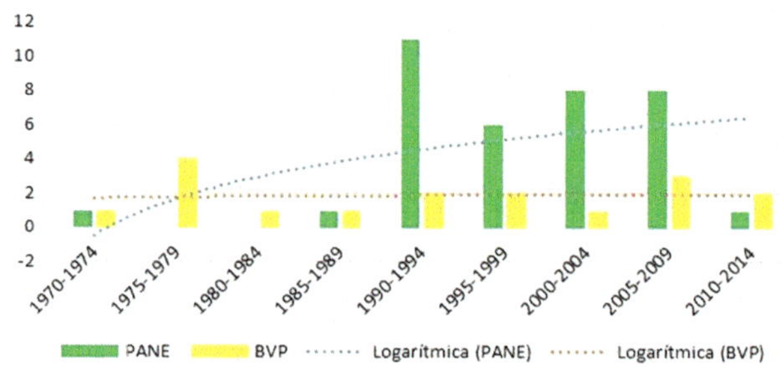

Fig. 5.4 Dynamics of *ANP-RAE* creation. (1970–2014)

Additionally, as this figure illustrates, the number of PANE and BVP created in the Amazon region, tends to stabilize over time, although the increase in the number of registered units (Fig. 5.3), contrasts with areas each time reduced (i.e., an inverse pattern). The contribution in land area is significant for *BVP* after 1985, although the biggest area and number of BVP created during the period of study, corresponds also to the five-year period of 1990–1994, with more than reduced, with the exception of the last five-year period, for the creation of a *BVP* with a significant extent half of the total area and around a third of the units today.

The expansion of extraction activity in the Yasuní national park, the core zone defined as Intangible Zone (ZI) for the uncontacted peoples Tagacri-Taromenane (ZITT) continued in August 2016 and revealed the dominance of petroleum politics over conservation and the protection of indigenous groups in voluntary isolation. Such prioritization

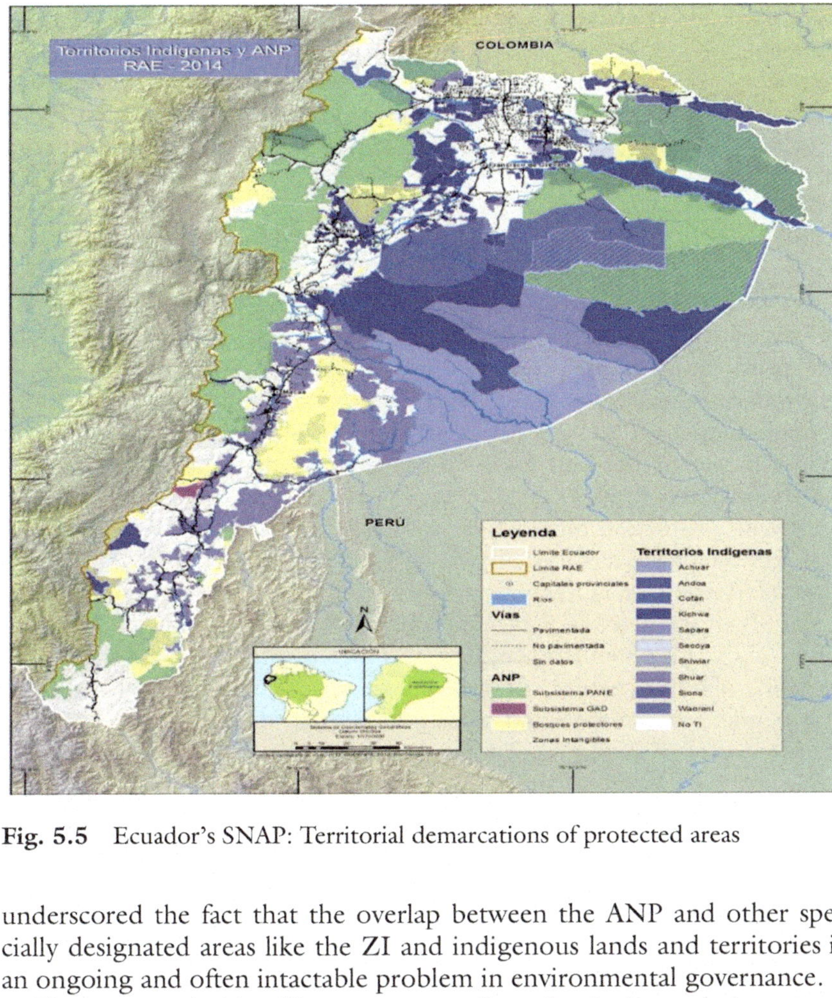

Fig. 5.5 Ecuador's SNAP: Territorial demarcations of protected areas

underscored the fact that the overlap between the ANP and other specially designated areas like the ZI and indigenous lands and territories is an ongoing and often intactable problem in environmental governance.

The case study identifies two types of overlap in its analysis: one of "simple" overlap between TI and one of the respective conservation units (PANE, BVP, ZI, or Socio-Bosque Program–PSB), and the other of "multiple overlap" where in addition to overlaps between indigenous territories and SNAP units are among other special category of conservation such as intangible zones (TI-PANE-ZI), or when a Socio-Bosque Program (PSB) is initiated within ZIs (López et al. 2016). It is key to note that four indigenous nationalities (all from the province of Pastaza), do not register areas of overlap with any kind of ANP in their respective territories (Achuar, Andoa, Sapara, and Shiwiar). In contrast, in the case of the Waorani nationality lands, there are areas of overlap with a PANE unit and ZI, so have considered its addition for the recent report cited here (Ibid), as seen below on Fig. 5.5).

Conclusions and Recommendations

Even with the normative legal and juridical framework in force for the protection of biodiversity and indigenous peoples, the TI and ANP are under threat from deforestation expansion, above all by concentrating forest resources "reserves" into geographically circumscribed zones. Extensive overlap between the ANP and TI, in particular, are a warning about the great need to improve governance mechanisms under these conditions, above all to encourage a territorial management in accordance with protection regulations, natural resource use, as well as for the full exercise of the territorial rights of these populations, with particularly careful attention to indigenous groups living in voluntary isolation.

Another set of critical mechanisms are those to monitor deforestation—above all when triggered by activities of economies of scale, integration of infrastructure, or projects of energy security—by means of independent assessments and of with unfiltered presentation to everyone involved in the TI and ANP. Current lands and territories of the RAE's indigenous nationalities share two features in common: (a) they are the foundation on which the sociocultural reproduction of those people depends, requiring protection of biodiversity and land-use restrictions; and (b) the fixed boundaries of communal control and/or individual and family use of the land or forests shape territorial management and planning for future use.

By settling inter and intra-communities boundaries, particularly between indigenous nationalities and other territorial groups, Ecuador's SNAP has also impacted the occupation and traditional management of forests and other ecosystems (i.e., wetlands or traditional horticulture systems, *chakra*). Due to accelerated population increases, sustainable management of those ecosystems is also critical.

Bibliography

Amazonian Network of Georeferenced Socio-Environmental Information (RAISG). 2012. *Amazonia Under Pressure*. São Paulo: RAISG.

——— 2015. *Deforestation in the Amazonia (1970–2013)*. São Paulo: RAISG.

De Marchi, M.S. Pappalardo, and F. Ferrarese. 2013. *Zona Intangible Tagaeri Taromenane (ZITT). ¿Una, Ninguna Cien mil? Delimitación cartográfica y pueblos indígenas aislados en el camaleónico sistema territorial del Yasuní*. Quito: CICAME-Fundación Labaca.

Dudley, Nigel (eds.). 2008. *Directrices para la aplicación de las categorías de gestión de áreas protegidas*. Suiza: Unión Internacional para la Conservación de la Naturaleza (UICN).

Fundación Amigos de la Naturaleza (FAN). 2015. *Atlas socioambiental de las tierras bajas y yungas de Bolivia*. Santa Cruz: FAN.

López, A.V., J. Ulloa, and J. Aragón. 2016. *Cartografía Histórica of las Áreas Naturales Protegidas y los Territorios Indígenas de la Amazonía Ecuatoriana*. EcoCiencia- Red Amazónica of Información Socioambiental Georreferenciada (RAISG), Quito. https://flacso.academia.edu/VictorLopezAcevedo.

López A. Víctor, M.O. Borja, and J. Aragón. 2015. Deforestation in the Ecuadorian Amazonia. Deforestation in Amazonia (1970–2013). São Paulo: Amazonian Network of Geo-referenced Socio-Environmental Information (RAISG).

López, A.V., F. Espíndola, J. Calles, and J. Ulloa. 2013. *Atlas Amazonía Ecuatoriana Bajo Presión*. Quito: EcoCiencia and Red Amazónica of Información Socioambiental Georreferenciada (RAISG).

Ministerio del Ambiente del Ecuador (MAE). 2007. *Políticas y Plan Estratégico del Sistema Nacional of Áreas Protegidas del Ecuador 2007–2016*. Quito: MAE.

Secretaría Nacional of Planificación (SENPLADES). 2013. *Plan Nacional del Buen Vivir 2013–2017*. Quito: SENPLADES.

CHAPTER 6

Colombia: Bridging the Gaps Between What Is Needed and What Actually Exists Regarding the Protection of Its Amazon

Ana María Hernández Salgar and Luz Marina Mantilla Cárdenas

Abstract This chapter examines how policy and its legal framework are crafted and applied, with a focus on the complex conditions of Colombia's Amazon region. The authors, two of the country's leading scientists and policy specialists, critique how law is formulated, how political interests shape its provisions, and how much regulatory and institutional support it actually enjoys. Colombia has long been recognized for its strong judiciary and its centralized administration, but the environment may be the area that benefits least from their support.

Translated by Judith Filc.

A.M.H. Salgar (✉)
Instituto de Investigación de Recursos Biológicos Alexander Von Humboldt, Bogotá, Colombia

L.M.M. Cárdenas
Instituto Amazónico de Investigaciones Científicas, Leticia, Colombia

M. Ungar (ed.), *The 21st Century Fight for the Amazon*,
https://doi.org/10.1007/978-3-319-56552-1_6

Keywords Amazon · Colombia · Environment · Indigenous peoples

Introduction

The seven million square kilometers of the Amazon basin are more than the world's lungs. They comprise a vibrant territory full of contrast, cultural expressions, social struggles, economic ventures, and yes, great biodiversity and ecosystem services stemming from the close relationship humans have built with their natural environment. This chapter seeks to highlight that due to its particular geographic features (such as its connections with the Andes to the West and the Orinoco region to the east), the Great Amazon region is one of the most challenging regions in the world to maintain.

The Colombian Amazon region occupies 42.42% of the country's territory (488,000 km^2) and 6.21% of the Amazon bioregion. While the numbers vary depending on the source, we can accurately state that this region contains more than ten of the country's 56 national protected areas (National System of Protected Areas). These areas are divided into the following categories: national natural parks, national natural reserves, and a flora and medicinal plant sanctuary.[1] So far, 1159 useful species have been identified that belong to 150 botanical families. In addition, scientists have found 674 bird species, 158 amphibian species, 195 reptile species, 212 mammal species, and 753 fish species (SINCHI 2016). The region's vegetation corresponds to the great biome of the tropical rainforest, within which are found many different types of biomes. The zonobiome represents 64.9% of the region; litobiomes (rock formations), 14.5%; helobiomes (areas that are greatly influenced by water, such as the alluvial plains of the Andes and Amazon rivers), 12%; peinobiomes (high plain areas, the Yari and La Fuga savannahs), 3.4%; and orobiomes (mountainous areas, the eastern slope of the eastern range), 4.7%. There are three subunits of orobiomes, namely, low, medium, and high mountains (see SINCHI 2007) (Fig. 6.1).

The Amazon basin, which stretches from the Andes foothills to the southern border of Colombia, ceased to be viewed as a structural unit as soon as Spain and Portugal divided it between them during the conquest and colonial periods. Later, during the height of rubber production and mining, local governments gave into private capital ambitions. As the region engaged in ongoing border disputes, such as the 1932 war

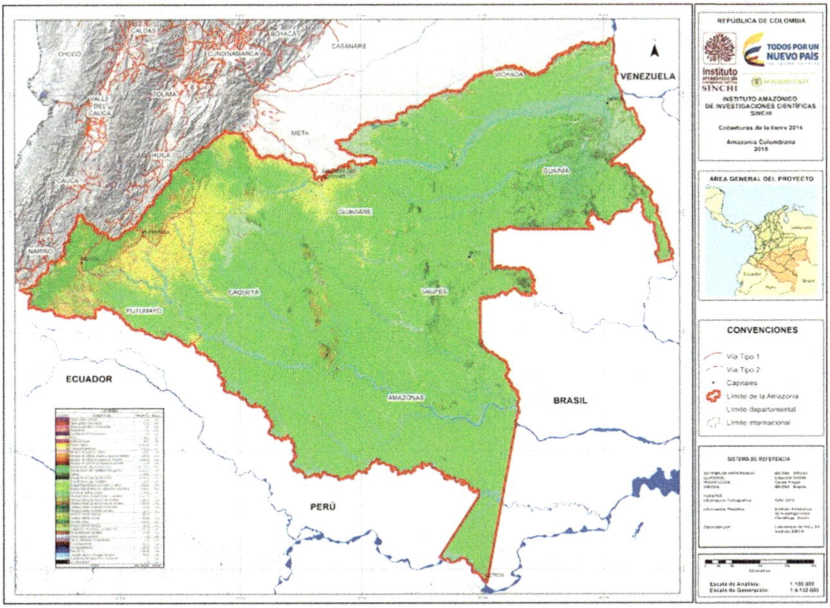

Fig. 6.1 Colombia's Amazon. *Source* SINCHI (2016)

between Colombia and Peru and illegal incursions from other countries, native populations were displaced and abused, leading to the deterioration of their culture and knowledge.

Until 1990, the Colombian Amazon region was organized into municipalities and a hamlet. This administrative typology, known as "national territories," did not change until 1991, when a new constitution was approved. Like other constitutions of the region, it affirmed the country's multiethnic and multicultural character. To help give that character concrete expression, Article 171 created two reserved seats in the Senate for Indigenous representatives elected in national districts. Such representation was strengthened by the 1993 Law 70 that created two reserved seats for Afro-Colombians in the Chamber of Deputies and Law 649 of 2001, which granted Indigenous representatives an additional seat in the Chamber. More pertinently for the Amazon, constitutional Article 297 allowed national territories to be elevated to the category of departments and Article 286 established new political divisions such as indigenous departments, municipalities, districts, and territories. These changes

Table 6.1 Extent of Colombia's Amazonia

Total surface area	483,164
Total population	960,239
Total indigenous population	86,417
Total Afro-Colombian population	28,016
Departments fully or partially located in the Amazon region: Amazonas, Caquetá, Guainía, Guaviare, Putumayo, Vaupés, Vichada, Meta, Cauca, and Nariño	10
Municipalities or hamlets fully or partially located in the Amazon region	78

Source Based on data from The System of Territorial Environmetal Information of the Colombian Amazon (SIAT: Sistema de Información Ambiental Territorial de la Amazonía Colombiana), www.siatac.co., SINCHI (2007) (based on the 2005 Census; DANE)

gave the Amazon region a stronger political position. Altogether, that region includes the entire Amazonas, Caquetá, Guaviare, Guainía, Putumayo, and Vaupés Departments and part of Nariño, Cauca, Meta, and Vichada Departments, reaching a total extent of 483,164 square kilometers. This recent history partly explains the strong influence of the central government on the Amazon's institutional organization, as well as the region's dependence on the central government (Table 6.1).

This vast territory, isolated politically for centuries and distant from the capital of Bogotá, had much closer connections with the nearby border areas of Venezuela, Brazil, Peru, and Ecuador. But it moved to the center of public attention during the administration of President Virgilio Barco (1986–1990). Barco's recognition of the environmental and cultural relevance of the Amazon region and, above all, of the ancestral right to land of the indigenous populations that inhabit it (more than 50% of the total Amazonian population) constituted a historic milestone. The Predio Putumayo indigenous reserve, the largest in Colombia with more than 6 million hectares, was created at that time. More than 140 indigenous reserves, protected areas,[2] the Amazon forest reserve,[3] and peasant farmer reserves and other entities were established later.

Even though these measures were a major advance toward land-use planning in the region, they have also generated conflict. The main trigger for such conflict is duplication: more than four million hectares (40,000 square kilometers or 15,444 square miles—the size of Maryland and Delaware combined) of overlap between protected areas, indigenous land, mining titles, watering districts, and titles for empty lots. As also discussed in the chapter on Ecuador, such overlap generates legal confusion, social conflict, and jurisdictional challenges.

An Economic, Sociocultural, and Environmental Mosaic

Spread over the vast Colombian Amazonia are 58 communities of just 1.5 million people, many of them indigenous, peasants, and Afro-descendants. Despite denunciations of the "white" government's neglect and lack of understanding of Amazonian cultural diversity, Colombia is very advanced in certain areas, particularly on legislative protection of indigenous peoples. Indigenous lands have been recognized both in the Constitution and in an array of laws. In addition, the land-use planning law gives indigenous peoples the option to create their own government with high levels of social autonomy and organization.

This legal foundation, though, has not fully prevented mistrust in the relationship between the federal government and the governments of the Amazon's indigenous peoples. Despite progressive law, in particular, is a lack of acknowledgment of their right to full self-governance when it impedes national economic and political interests. The resulting distrust compounds the long-running difficulty of developing a fluid dialogue that includes local authorities, regional indigenous associations, and national organizations, along with the other actors who should be engaged in the multiscale process of policy development. Further complicating these relationships is the fact that indigenous peoples are not the only inhabitants of the Amazon region, and the legislation aimed at protecting them is often not applied to or supported by the other communities. Decision-makers for the Amazon must therefore listen to, acknowledge, and respect the region's perspective to a greater extent. They should start locally and then move on to the national and international levels. When we look at the region from the inside out, we can understand the nuances that shape its dynamics. To promote its development, we must first meet its basic needs, particularly since the region's growing population requires a healthy environment. These needs must be taken into account when formulating sustainable development policy for the region. Demographics, environmental issues, and poverty eradication must be integrated into all policies, plans, and programs.

Part of this policy integration is the curtailment of unsustainable modes of production and consumption, as well as the negative effects of demographic factors—from deforestation to pollution (Salazar and Riaño

2016). Identity problems also need to be solved. While indigenous peoples have been recognized, there are many other communities and residents who have lost their identities or have adopted the Andean lifestyle. The resulting degradation of the region's diverse cultural heritage adds to the ongoing tension between centrist development models aligned with national sector interests, and local development models focused on the local community. Both models make use of the region's biodiversity and ecosystem services, but in conflicting ways. Despite understanding the environmental and social significance of the region at the global level (its contribution to the planet's climate and water security), the country's growth model still applies unsustainable exploitation strategies. This situation is reflected in the region's current conditions; according to SINCHI, 80% of the Amazonian forest is well conserved, in contrast with the changes suffered by more than 10% of the forest cover in the last 12 years. Incentives for the development of financial ventures that are foreign to the regional reality and do not take into consideration the Amazon's biodiversity, leading to activities with highly negative impacts on its ecosystems. Among those activities are the increasingly large cattle industry; extesive logging; a sharp rise in illegal mining and illicit crops; and the introduction of invasive exotic species (more than 250 were reported by SINCHI).

Such an increase must be tackled not only because of the social problems it generates, but it is also a threat to the regional biome. In the Amazon, the great diversity of species being used for the local population's sustenance is at odds with the great risk of poverty posed by the loss of those species to illegal trade. For instance, despite the abundance of plants such as Rubiaceae, Melastomataceae, and Arecaceae, the number of endangered species is rapidly climbing, according to UICN criteria. Among those being harvested commercially are Colombian mahogany, mahogany, cedar, rosewood, and the Andaqui cinnamon tree (Cárdenas et al. 2015). Despite the existence of forestry roundtables, a range of environmental authorities, and both governmental and non-governmental watchdog organizations, illegal exploitation of these species has reached a critical danger status.

So even though Colombia has signed international treaties and adopted national regulations to implement them—such as with the Convention on International Trade in Endangered Species of Wild

Fauna and Flora (CITES)—several questions are still worth asking. To what extent is environmental legislation being enforced, and what real power do authorities have to safeguard the country's natural assets? Or are authorities aware of the illegal activities but do not consider them relevant or allow them to continue because they attach more significance to the economic benefits or because they are distracted by the armed conflicts? Such questions pertain to a situation that is not exclusive to Amazonian flora. With regard to wild fauna, illegal trade directly influences the food security of local communities, as evidenced by inventories conducted in Vaupés Department and in little-known areas such as Tiquié. These inventories show that if the community loses its source of food, the risk to its members' lives becomes so great that the whole community can disappear. For this reason, many local inhabitants who belong to indigenous communities want to continue monitoring species that have been identified as vital to them (Osorno et al. 2014). Unfortunately, however, rather than understanding the region's own dynamics, needs, and circumstances, the government and external organizations insist on solving the illegal trade of fauna by importing foreign models that are more adequate for contexts like Africa's. In the case of bush meat, for instance, are proposals such as the establishment of use quotas to promote commercialization that will generate financial resources for local communities—which may work in Africa but are not appropriate for the Amazon.

The region's endogenous capacities must be strengthened so that it can generate its own development drivers based on environmental conditions and local cultural and social priorities. Illegal markets, especially the speculative land market, must be eliminated, for they push forward the colonization frontier and jeopardize ecosystems within and without protected areas. According to SINCHI, "management decentralization, the new legal rights of minorities and ethnic groups, the rise in mining activities, cross-border integration, and the growth of illegal crops and its manifold consequences, among others, have led to the development of urbanized migrant communities along the region's highways and water corridors in a pattern often known as a settlement ring.

As discussed by Arcila (2010), despite the fact that the area's great biodiversity has generated (legal and illegal) trade of flora and fauna species, the population is highly dependent on extra-regional markets (at the national and even the international levels, due to its border status), with products traded in centers that are accessible to the population. In addition, extensive economic activities in situ have led to soil depletion and diminished resources, prompting waves of migration toward metropolitan areas. Seen from the perspective of basic needs, population and urban growth require greater road and public service development, and therefore create financial, political, and social pressure that then increases poverty and unemployment. But rather than thinking about conservation, national and regional policymakers seem to respond to the pressing need to generate income through traditional production systems, which they believe is the best or only way to solve critical poverty, education, and health problems. In the process, they overlook opportunities based on knowledge of the region's potential, and which could be profitable. While there is an impulse to create local markets and bio-commerce networks, there are few financial incentives and market niches, and it is hard to gain access to production processes with adequate standards. As a result, the sustainable exploitation of biodiversity does not appeal to local residents, save for tourism, and it is much cheaper to acquire goods and services that are produced in other areas of Colombia or in other countries.

Not only are decisions on the region imbalanced in economic or social terms; in some cases, the conservation discourse overlooks the reality faced by the community in its struggle to meet basic survival needs. Public policy instead needs to more comprehensively develop solutions that acknowledge that the best way to ensure social welfare is the adequate management of biodiversity and ecosystem services. Their loss or damage will generate more poverty, disease, and cultural collapse. To formulate the right policy, policymakers should engage in a dialogue to tackle problems and find solutions across sectors. While there have been efforts in that direction through dialogues on long-term solutions, such as collaboration among agencies and regional roundtables, many of them are dialogues of the deaf. This is also true for the environmental sector, which is often stuck in conflicts over political or legal agendas. An example of this recurring problem is the overlap of land management mandates and local jurisdiction, as mentioned at the beginning of this chapter, which impedes the strengthening of a comprehensive governance of the region because agencies rarely account for or accept the existence and contributions of other forms of governance already present in the region.

Environmental Institutions and State Enforcement

To fully understand this scenario, it is important to examine existing governance institutions, since they serve as a platform to support organization processes on environmental issues in the Amazon. Thanks to the concerns raised by conservation movements and the 1972 Stockholm Summit, the world regards the Amazon region as critical to the planet's climate, with its protection taken up by the Amazon Cooperation Treaty Organization in its focus on cross-border cooperation. Thanks to this organization, by the time the 1992 Earth Summit was held in Rio de Janeiro, an Amazonian political bloc of common interests had taken shape to make the region's conservation and sustainable development as a priority in the global agenda.

At the same time, the border agenda for Brazil, Venezuela, Ecuador, and Peru, developed in the framework of border agreements and related neighbor relations commissions, extends governments' responsibilities to the implementation of practical policies in a region that may be geodesically divided but has a dynamic social, cultural, economic, and environmental life. The efforts undertaken in international negotiation spaces and the strengthening of environmental policy in Colombia (whose constitution reflects the need to develop an agenda and a suitable organization to tackle environmental responsibilities because it is one of the most privileged countries in terms of water wealth and biodiversity) have increased the relevance of the Amazon region in the country.

However, as was pointed out earlier, from a legal and policy perspective, this vision is antiquated and needs to be adjusted to the actual situation in the region. There are areas in Colombia with problems that are common to all Amazonian countries, such as mining, illegal logging, overfishing, and illegal crops. Dealing with these issues requires the creation of legally binding agreements that will not affect the countries' sovereignty but will facilitate coordinated actions to protect the region and its inhabitants. Some programs have already been developed in the context of the overall Basin, such as the Regional Fund for the Guyana Shield Facility. With the participation of six countries, the Fund aims to develop methodologies and projects to protect and use its biodiversity and ecosystem services in a sustainable way, promoting, in turn, development and social welfare projects.

At the same time, Colombian civil society is promoting complementary and more focused initiatives such as the Path of the Anacondas program. That program's goal is to consolidate an ecological-cultural mosaic that will make it possible to maintain the connectivity of the Andes-Amazon-Atlantic ecosystems, and thus ensure the environmental services of the Amazon's tributaries and seek innovative answers to climate change. Additional projects, conducted in the framework of the OTCA, also work to promote joint cooperation in order to improve environmental health, restore water resources, or conserve biodiversity. Nonetheless, both at the regional and national levels, each project or program is disconnected from the others, and so its contribution to the comprehensive management of the region is fragmentary. This issue needs to be solved urgently.

"The definition of the scope, instruments, and territorial differentiations in the promotion of conservation makes it necessary to develop a major land-use program that will take into account sub-regional physical differences and populations" (CEPAL and Patrimonio Natural 2013). Yet this project will not be viable without an institutional structure and policies that support the actions required to solve the problems that have already been identified in the region in general and in the Colombian Amazon region in particular. This section specifically addresses such a need from an environmental perspective.

While the National Natural Resource and Environmental Institute has existed since 1968 and Executive Order 2811 of 1974 established the National Code for Renewable Natural Resources and the Environment, it was not until 1993 that the Colombian environmental institutions were strengthened with the passage of Law 99. This law established the National Environmental System (SINA), which, thanks to the Amazon region's strategic significance, recognizes the activity of the Colombian Corporation for the Amazon Region, Araracuara, and integrates it into the system as the Amazonian Institute for Scientific Research, SINCHI. SINCHI's goal is to conduct, coordinate, and disseminate high-level studies and scientific research focused on the Amazon region's biological, social, and ecological reality (Executive Order 1603 from 1994).

The other scientific institutes that were established to support the ministry's environmental administration are the Institute of Hydrology, Meteorology, and Environmental Studies, IDEAM, the Institute of Marine and Coastal Research, Invemar, the Institute of Research on the Pacific, IIAP (which conducts research on the biodiversity of the biogeographical Chocó), and the Alexander von Humboldt Institute (in charge of coordinating scientific research on the biodiversity of the Colombian mainland). Given that their goal is to produce and manage scientific knowledge needed to inform decision-making related to the comprehensive administration of biodiversity and ecosystem services, the joint work of SINA and institutes such as SINCHI and Humboldt is key to development of policies and programs that are suitable for the country and the region. SINA's formation SINA has enabled a state that has been traditionally recognized for its highly centralist governments to generate and implement policies with a regional approach. In this way, while Bogota has developed standards and established jurisprudence, it is in the territory itself where institutions, communities, civil society, academia, and the private sector are being strengthened in order to tailor environmental regulations to local situations and needs. The Special Administrative Unit for National Natural Parks and its Amazon Office, regional autonomous corporations, and urban environmental authorities, which are also part of SINA, play a key role in this decentralization of the environmental sector.

Nevertheless, as discussed above, the creation of this regional governance structure also generates political tension between central and regional authorities, as well as among different forces that have political interests in the region. In the Amazon, that set of forces is complex. On one side are the traditional native inhabitants (whose views on the region have not been fully recognized); the settlers fighting for land and beliefs; the tourists, who generate income; private businesses in search of new market opportunities; and government agencies, civil society organizations, and the academy, which invest at the local level. On the other side are those who engage in illegal logging and wildlife trade, drug traffickers, among others, who of course do not comply with environmental policy. Such organized criminality, as discussed throughout this book, is a major and growing challenge for the entire Basin. But Colombia's Amazon faces the additional complication of its still-simmering armed conflict. In October 2016, a peace deal that was years in the making was rejected in a referendum. Although both the government and FARC are determined to pick up the pieces, the rejection shows how unsettled the conflict remains. In rural regions like the Amazon, part of that uncertainly is the continuing influence and presence of guerrillas, and the paramilitary units. Altogether, these overlapping conflicts created constant tension in the Amazon. Combined with the region's high levels of poverty and other social-economic ills, such conditions require the

Colombian government urgently embrace its Amazonia as a priority in order to protect its cultural and natural wealth and diversity.

Such uncertainty makes it even harder for global or national policy to crystallize at the local level. Moreover, there is a gap between the standards enshrined in legislation and their application through clear, transparent and, above all, effective policy that will stop environmental deterioration. If this gap did not exist, Colombia would be a leader in the fulfillment of biodiversity conservation goals, such as the Aichi Goals in the framework of the Biology Diversity Agreement (CDB). These guidelines provide a new framework that has produced a novel cross-cutting vision on the need to integrate biodiversity and ecosystem services into the country's structural planning. These guidelines were part of the National Policy for Comprehensive Management of Biodiversity and Ecosystem Services, PNGIBSE (Ministry of Environmental Affairs and Sustainable Development, MADS, 2012), which set out rules on conservation, sustainable use, and knowledge of natural resources from a cross-cutting territorial perspective so as to foster the strengthening of the regions by developing specific biodiversity action. The effectiveness of its implementation will be measured by way of the recently launched National Action Plan on Biodiversity (PAB). This plan contains indicators that will show whether the country will still be unable to comply with its environmental obligations by 2025, or whether it will have started to improve its performance, more specifically, in relation to biodiversity, from the local to the global level.

Colombia's Fifth Report to the Convention on Biological Diversity (CDB) in 2014 showed that there is awareness of the alarming reduction of national biodiversity, which is due to both natural and human causes. The latter are primarily linked to illegal logging, the overexploitation of the soil and wildlife, and the expansion of the agricultural frontier. With the main drivers of loss of biodiversity identified and new policy in place, in order to reach the much longed-for institutional and legal consistency, land-use plans must take into account financial as well as environmental variables. Spaces for dialogue, such as the ones created to draft the Agenda 21 Amazonia,[4] can reappear in the context of the regional implementation of the National Biodiversity Plan (PAB) by way of territorial agreements. Concurrently, it is imperative to implement policies for comprehensive border resource management with the participation of neighbor relations commissions.

Unlike many who consider the environment an obstacle to development, we believe that biodiversity and its ecosystem services offer the best opportunities for growth in the region. Given the ongoing administrative inertia and economic resistance, though, we must conduct in-depth and carefully presented strategies to induce markets to adopt non-timber forest products. It is also necessary to strengthen local communities so that they can take on the challenges posed by natural resource management with the value added and the quality required by the demand for these products. For the region to become profitable again, affected areas must be restored and sustainable production systems must be developed with the help of inter-sectoral agreements.

In addition to economic development, biodiversity can be a motivating factor in facilitating the resolution of many of the policy inconsistencies that loom over the region. At the international level, in the context of the CDB or the Framework Convention on Climate Change, Colombia has decided to reduce both the loss of biodiversity and the production of greenhouse gases. In addition, by fulfilling commitments such as the Bonn Challenge, the country hopes to reach approximately 1 million hectares of restored biodiversity in 4 years. With the help of the international community, funding was secured in the Amazon region to promote a "zero deforestation" policy for 2020.

Of the 120,000 hectares that are cut down each year in Colombia, 58% is in the Amazon region and extends into departments such as Caquetá, Guaviare, and Putumayo. But by launching vigorous new strategies like Visión Amazonia,[5] the Ministry of the Environment and Sustainable Development expects to solve the regional problem by means of national policies. This problem is not just environmental; it is generated by and affects a variety of sectors and actors. Yet, while these policies are based on international commitments, have been formulated by the administration as a whole, and affect the entire public and private institutional structure in the Amazon region, the environmental sector is considered solely responsible for their implementation. For this reason, the government drafts policies that are ignored by all the other sectors.

Opportunities are thus thwarted from within, aggravated by a lack of communication and method in the formulation of public policy for the Amazon region. Interests respond to plans that foster development processes based on standards that take neither the value of biodiversity and ecosystem services nor the aspirations of the local society into account.

Hopes to get past this policy quagmire has increased with the shift in the country's political juncture, in part from the current de-escalation of the country's long-running armed conflict, mainly between the government and the Revolutionary Armed Forces of Colombia (known commonly as FARC: *Fuerzas Armadas Revolucionarias de Colombia*), which lasted more than 50 years. Looking toward the future, the current administration is hoping to join the Organization for Economic Cooperation and Development (OECD). In this context, the government adopted a 2014–2018 national development plan, entitled *Everyone for a New Colombia*, based on six cross-cutting strategies: competitiveness and strategic infrastructure, social mobility, transformation of the countryside, safety, justice, and democracy for the construction of peace, good government, and green growth. It is refreshing to see the inclusion of sustainable development, at least on paper. In order to facilitate its implementation, the plan was "regionalized" in order to differentiate and prioritize issues for each of the country's new geographic "divisions." In addition to what has been historically known as the Amazon region, this approach will mean that national planning will invest in what is currently called the Center-South-Amazon Region, which comprises Andean departments like Tolima and Huila as well as the traditional departments of Caquetá, Putumayo, and Amazonas. In this large area, focus will be on conservation and sustainable use of the cultural and environmental assets; development of the region's agricultural potential in a sustainable way; and a shift toward greener businesses in the hydrocarbon, mining, and farming sectors.

It is still too early to tell what the outcome of this national project will be. We do not know yet if more chaos will ensue due to the overlap of biophysical and economic regionalization and political regionalization or if regional construction dialogues will take root. The truth is that in order to achieve a new Colombia and a new Amazon region, authorities must be entirely uncontaminated by political interests and must comply with their obligations. Colombia has a great opportunity to makes its Amazon a laboratory for peace and sustainable development.

Notes

1. The difference may be due to the way the boundaries of the Amazon region are drawn. According to the National Natural Parks' Special Management Unit, the region has eight parks, two reserves, and one sanctuary. Other sources add the parks adjacent to the Orinoco region and the Andean foothills.
2. They include Yaigojé Apaporis, Serranía de los Churumbelos, Río Puré, La Paya, Cahuinarí, Alto Fragua Indi Wasi, Amacayacu, and Serranía de Chiribiquete National Parks, Puinawai and Nukak National Natural Reserves, and the Orito Ingi-Ande Flora and Medicinal Plant Sanctuary.
3. Its boundaries go from Santa Rosa de Sucumbíos on the Ecuador border toward the northeast up to the mountain in Picos de la Fragua; from there, following a line 20 km west of the Eastern Range, up to the Oseras Heights; from there, toward the Ariari River, and following this river up to its confluence with the Guayabero or the Guaviare River, and downstream to the mouth of the Orinoco. Then the boundaries coincide with the border with Venezuela and Brazil up to the Amazon River.
4. The Agenda 21 for the Colombian Amazon was developed under SINCHI to coordinate a joint line of action between the government and civil society based on a sustainability approach regarding ecological, economic, social, and political processes and to improve its population's quality of life and dignity.
5. Promoted by the Ministry of Environmental Affairs and Sustainable Development, the Visión Amazonia 2020 program seeks to reduce deforestation to zero in this region. The program was developed in the framework of the REDD+ National Strategy. Among its objectives are the expansion of the Chiribiquete National Park and the consolidation of its buffer zone in the context of the regional program of sustainable landscapes (Brazil, Colombia, and Peru). This program, which has been approved by the Global Environment Facility, aims to maintain 73 hectares of forests and promote the sustainable management of the region. At the same time, it will allow the reduction of carbon emissions by 2030.

Bibliography

Arcila Niño, O.H. 2010. *La Amazonia colombiana urbanizada: un análisis de sus asentamientos humanos* [Urban Development in the Colombian Amazon Region: An Analysis of the Region's Human Settlements]. Bogota: Instituto Amazónico de Investigaciones Científicas, SINCHI.

Cárdenas López, D., N. Castaño Arboleda, S. Sua Tunjano, L. Quintero Barrera et al. 2015. *Planes de Manejo para la Conservación de Abarco,*

Caoba, Cedro, Palorosa, y Canelo de los Andaquíes [Management Plans to Conserve Colombian Mahogany, Mahogany, Cedar, Rosewood, and Andaqui Cinnamon Trees]. Bogota: Instituto Amazónico de Investigaciones Científicas, SINCHI.

CEPAL (Economic Commission for Latin America and the Caribbean) and Patrimonio Natural. 2013. *Amazonia posible y sostenible.* Bogotá: Cepal an Patrimonio Natural.

Instituto SINCHI. 2007. *Annual Assessment of the Status of the Ecosystems and Environment of the Colombian Amazon Region. 2006* Leticia: SINCHI.

Instituto SINCHI. 2016. *Herbario Amazónico Colombiano* [Colombian Amazon Herbarium]. http://sinchi.org.co/coleccionesbiologicas/index.php. Accessed May 2016.

Osorno Muñoz, M., N. Atuesta Dimian, L.F. Jaramillo Hurtado, S. Sua Tunjano, A. Barona Colmenares, and N. Roncancio Duque. 2014. *La despensa del Tiquié: Diagnóstico y manejo comunitario de la fauna de consumo en la Guayana colombiana* [The Tiquié Breadbasket: Community Assessment and Management of Consumption Fauna in the Colombian Guyana]. Bogota: Instituto Amazónico de Investigaciones Científicas, SINCHI.

Salazar Cardona, C., and E. Riaño Umbarila. 2016. *Perfiles urbanos en la Amazonia colombiana, 2015* [Urban Profiles in the Colombian Amazon Region, 2015]. Bogota: Instituto Amazónico de Investigaciones Científicas, SINCHI.

CHAPTER 7

Environmental Penal Control in Venezuela: Amazonia and the Orinoco Mining Arc

Manuel Joel Díaz Capdevilla

Abstract An assessment of the ability of civil society, the media, and other forms of vertical accountability to enforce environmental law is the focus of this chapter. Although Venezuela has one of the world's strongest sets of legal protections, enforcement is undermined by the country's political strife and economic crisis. The potential for local groups, independent organizations, and community efforts to make up for these problems is also examined.

Keywords Amazon · Environment · Indigenous peoples · Mining · Rainforests · Venezuela

Translated by Mark Ungar.

M.J.D. Capdevilla (✉)
Primer Canal Ecológico de Venezuela, ECOVISION, Caracas, Venezuela

M. Ungar (ed.), *The 21st Century Fight for the Amazon*,
https://doi.org/10.1007/978-3-319-56552-1_7

Introduction

Human settlement in areas that cannot sustain it leads to damaging disruptions of natural cycles that, in turn, reverberate against such settlement. Focusing on the case of Venezuela and its Amazonia region, this chapter illustrates such a vicious cycle on a national scale. It begins by describing the country's extensive environmental institutional apparatus, created when the national government was intent on protecting Venezuela's vast natural resources by reducing their vulnerability to entrenched patterns of exploitation. It then examines how well those protections address the causes of the country's many forms of air, soil, and water pollution; how they balance the environment with economic development; and whether public and private efforts—such as the management of aquifer, mining, and agricultural resources—will ensure future generations a healthy environment and better living conditions.

To bring all these challenges together, this chapter focuses on the Orinoco Mining Arc, located in the states of Amazonas and Bolívar. With Decree N° 2.248, which went into force in July 2016, the government launched a massive exploitation of that area, which will proliferate already mounting problems such as the spread of contaminants like mercury and lead; indiscriminate deforestation of one of the continent's great forest ecosystems; and irreparable damage to the area's biodiversity, critical resources such as water, and the indigenous ethnic groups inhabiting the area. The deterioration of Venezuela's environment has been, in recent years, an issue of primordial importance for First World countries. Ironically, in Venezuela itself, which is classified as one of Latin America's six "mega diverse" countries and considered among the ten world priorities for the conservation of biodiversity, the debate over the environment has not had the same vigor. As seen in the Orinoco Mining Arc, the government has used the law against nature with no regard for the consequences. As a conclusion, this chapter provides recommendations on how to best move forward.

Venezuela's Environmental Structure

Before such policies emerged, Venezuela was a pioneer in the establishment of environmental institutions. Basic citizen rights are fundamental for a democratic regime, to which the concept of a right to a healthy environment has been added through growing environmentalism over the last several decades in Venezuela, reaching full recognition in the country's 1999 Constitution, which consecrates in a novel and advanced approach the state's obligation to protect the environment (Articles 127, 128, and 129). Upholding this new right has given an enormous responsibility to judicial officials, who must balance individual and collective rights. To support that effort, Venezuela was the first country in South America to create an Environmental Ministry and to formulate a model environmental legislation. Much of that legislation focuses on the enclaves designated as protected zones to protect the country's diverse geography, from coastal dunes and deltas to lakes and plains. Many of those zones are concentrated in the Amazon.

Together, this foundation paved the way for the establishment in the Attorney General's Office (*Fiscalía Nacional del Ministerio Público*) of the Office of Integral Defense of the Environment and Environmental Crime. This unit is empowered to investigate all environmental crimes and strengthen both the training of and legal controls by prosecutors (*fiscales*), as well as of officials of all state security agencies. Together, these state entities are obligated "to procure environmental protection as a common good ... through the exercise of penal and civil action to establish culpability and sanctions that correspond to individuals, public officials and juridical persons, as well as prevention that contributes to sustainable ecological, social and economic development." These actions then continue into the courts within the Special Environmental Penal Jurisdiction, established in the 2007 Environmental Law (Articles 136–137) and comprised of courts authorized to handle environmental actions or omissions considered to be criminal acts, in all areas of the law. Many of these enforcement powers are detailed in the 2012 Environmental Penal Law, which gives the statewide leeway against environmental crime.

In practice, though, a range of obstacles cripple this carefully constructed legal and administrative structure. The most glaring is the fact that, as in Bolivia, the environmental courts have not actually been established, leaving environment criminal cases in the regular penal courts. The already over-worked judges in those courts must then grasp the role and nature of environmental resources in their areas (such as fauna, flora, vegetation, demographics, protected areas, and soil) and the many laws that govern them in Venezuela, such as the Law of Fishing and Aquatic Spaces, Law of Forests and Forestry Management, Wildlife

Protection Law, Water Law, and the Law of Biodiversity Management. Another set of obstacles are the constant delays and obstruction in the criminal justice process. Impunity is particularly rife, with very little help from the *Fiscalía General* (Attorney General), for which such elimination of impunity is officially its primary job. Another delay is of efforts to establish rigorous mechanisms—such as time frames and deadlines—to move cases to the Supreme Court (*Tribunal Supremo de Justicia*). As discussed in this book's introduction, environmental law needs judicial guidance, which cannot happen without judges to provide it. An even more glaring disappointment in Venezuela is of state agencies responsible for holding the state accountable, particularly those within the country's "Citizen Power" branch that was created with much revolutionary fanfare in the 1999 Constitution. Two such agencies are the *Contraloría General*, which supposedly works to ensure a transparent state administration, and the *Defensoría del Pueblo* (national ombudsman), which was created for the sole purpose of receiving and handling citizen complaints over state officials and actions. In countries like Bolivia, as also discussed in the introduction, the ombudsman has taken a role in the environmental protection.

Venezuela further loosens itself from compliance with environmental protection through its non-participation in the Clean Development Mechanism (CDM) of the United Nations Framework Convention on Climate Change (UNFCCC), the international framework to address climate change. The CDM is one of the Flexible Mechanisms of the 1997 Kyoto Protocol to facilitate emissions reduction projects that generate Certified Emission Reduction units (CERs), which in turn may be traded in emissions trading schemes. CDMs were designed to assist country signatories in Annex I, which are those who commit to reductions in greenhouse gases, and to help those not included in Annex I to move toward sustainable development. Along with the rest of Latin America, Venezuela is a non-Annex party, which means it signed the treaty but without binding targets. Unlike the rest of Latin America, though, Venezuela has been one of the world's biggest oil producers, and thus a far greater contributor to climate change than the other countries. As discussed below and in the other chapters as well—such as with the expansion of mining throughout the Amazon—this stance reflects a continuing reliance on extraction.

As a federal state, environmental enforcement is further hindered by the always politicized relations between state and federal levels. Through its National Environmental Authority, the federal state is authorized to exercise control over activities likely to degrade the environment, but only without infringing on the jurisdiction of the states, municipalities, peoples, and indigenous communities. Venezuela's long-domineering federal executive, though, has always kept those relations imbalanced, since presidents have not hesitated to use their considerable financial and political weight to impose policies and officials on the states. When the executive does back off, though, the lack of political will, finances, and institutional capacity in conservationist management by regional and municipal agencies cannot take up the slack, hindering protection. Poor coordination among environmental and development programs further loosens regional ties between conservation and sustainable development efforts.

A final set of obstacles, finally, is within society. On one level, there is limited and restricted public access to environmental information, particularly the EIAs of large and medium-scale development projects. More broadly lacking, though, are citizen education, consciousness, and knowledge of environmental regulations—together limiting their ability to organize, act, and pressure the government. Compounding these weaknesses is anemic inter-institutional coordination among environmental non-governmental organizations (NGOs) at all levels.

The Orinoco Mining Arc: A Swath of Destruction

To illustrate how all these obstacles develop within their larger context of global economics and extraction, we turn to Venezuela's Amazon and the Orinoco Mining Arc decree. Despite its rhetoric about the need to advance toward a sustainable economy, the national government decreed the "Exploitation of the Mining Arc of the Orinoco" that designated 111846.70 km^2 of national territory in the southern margin of the Orinoco (12.2% of the national territory) with an open invitation to transnational businesses. The area contains around 200 million tons of bauxite and some 44 million tons of gold and diamonds, and investment contracts have been signed for a total of US$ 4.5 million for 2016 and 2017. Such mining—particularly on the extraordinary scale planned in the Arc—embodies a short-term and short-sighted sprint for quick income at the cost of irreversible socio-environmental destruction of a significant proportion of national territory. It also poses an ethnocidal threat to the nine indigenous groups who live in the area, as discussed below, none of whom was ever consulted and their objections to large-scale mining were completely ignored, trampling over their constitutional rights to be fully consulted and sufficiently informed of all public policies that may impact their lives. The decree, in fact, was initiated solely through a presidential decision, absent any public debate in a society defined by its own constitution as "democratic, participatory, popular, multi-ethnic and pluri-cultural." Not even the larger population was provided with basic information; contracts were signed with no public comprehensive analysis of how those monies are earmarked, and how much it is going to cost current and future generations.

But economics, as discussed throughout this book, trumped rights. This massive mining endeavor is one of the 14 priority policies to strengthen Venezuela's moribund productive apparatus as part of the Bolivarian Economic Agenda, in conjunction with the nation's Economic and Social Development Plan (Article 4). Aiming to finally break free of dependence on petroleum, the plan is based on the government's assertion that mining is one of the most immediately accessible sources with which it can revive the economy following the collapse in the global price of crude, which triggered a drop in national income of over 90% in several years since 2013. But far from representing a strategic alternative vision to a rentier logic that has been predominant in Venezuela for the past century, in which the state derives its income almost entirely from extracted resources, this decree instead is a doubling down of extractivism and thus only deepens the country's rentier economy (Fig. 7.1).

The impact on forest and water resources will be tremendous. Most immediately, the project endangers huge parts of Bolívar and Amazonas states and their natural assets, such as the Caroní, Caura, Paraguay rivers; the Forest Reserve of Imataca; the world-famous Canaima National Park, which comprises 3 million hectares in the southeast of the country and is the primary home of the massive flat-topped mountains known as *tepuis*; and Angel Falls, the world's highest uninterrupted waterfall. The Arc's nearly 112,000 km^2 also includes humid tropical forests, great extensions of Savannah with fragile topsoil, an extraordinary biodiversity, critical sources of water, and indigenous people's territories—all of which are threatened if the decree's intended activities are carried out. Despite advances in reducing extractive technology's carbon footprint, mostly in the petroleum sector, no existing technology for large-scale mining is compatible with environmental preservation. International experiences in

Fig. 7.1 Political map of Venezuela, with the Orinoco Mining Arc area shaded in *gray*

this regard are overwhelmingly conclusive. In forested regions, such as a large proportion of the Mining Arc, large-scale and open-sky mining unavoidably involves massive and irreversible processes of deforestation. The zone's great wealth of biodiversity would be severely impacted, generating the loss of numerous species. Since the Amazon is a vital defense against climate change, such deforestation means an increase in greenhouse gases and a simultaneous reduction in the capacity of those forests to absorb and retain those gases—thus accelerating global warming. Instead of prioritizing the urgent need to halt the deforestation that has already been triggered by illegal gold mining, timber, and cattle ranching in the region, the legalization and promotion of large-scale mining activities planned in the Orinoco Mining Arc zone will trigger a rapid acceleration of these mutually reinforcing deforestation dynamics.

The mineral whose extraction has been most strongly promoted by the government is gold. According to estimates by the Ministry of Oil and Mining and by Eulogio Del Pino, President of the national oil company PDVSA (Petróleos de Venezuela), the zone's auriferous reserves are approximately 7000 tons, which at current prices is worth some US$280 billion.[1] Gold mining uses agents and chemicals, particularly cyanide and mercury, which are highly contaminating of soil and water. Aquatic life in the zone's rivers, in fact, have been contaminated by mining activity for the past several decades, placing at risk the health of the zone's inhabitants as well as those who live downriver. A large proportion of the region's current population has elevated levels of mercury in their systems, which the Mining Arc project would of course only multiply.

Open-air exploitation of gold requires excavation of immense volumes of soil for each unit of gold extracted. New open-air mining technologies make it profitable for business when the gold is found to be in concentrations up to 1 gram for each ton of dug-up dirt; that is, a relation of a million to one. To maintain that ratio, extraction in the next 70 years of these 7000 tons of gold, as announced by the government, would require removing some 7 billion tons of earth—gigantic mountains contaminated with arsenic and other toxins, which would dramatically and irreversibly alter the entire regional environment. Once these immense volumes of material are removed and crushed, the most efficient method to extract the gold is lixiviation with cyanide, which itself requires great toxic lakes that if not waterproofed will contaminate the ground and subterranean waters. As seen around the world, these lakes can and do flood their banks, with catastrophic consequences. Any claim that such land can be returned to its original state is, at the very least, disingenuous or dangerously delusional.

This decree is a blatant violation of the environmental rights and responsibilities expressly established by the Constitution, environmental legislation, and international agreements to which Venezuela is a signatory, such as the Biodiversity Convention. As Constitutional Article 127 states, "It is a right and an obligation of each generation to protect and maintain the environment for its own benefit as well as of the future. Every person has the individual and collective right to enjoy a secure, health, and ecologically balanced life and environment. The state will protect the environment, biological and genetic diversity, ecological processes, national parks and natural monuments and all other areas of special ecological importance [...]." All those rights, clearly, are completely disregarded by the policies articulated in the Orinoco Arc decree. In the first place, contrary to Constitutional regulations, territory opened for mining exploitation includes Areas under a Special Administration (ABRAE: *Áreas Bajo Régimen de Administration Especial*), National Parks, Forest Reserves like Imataca, Natural Monuments, and sacred places for indigenous peoples—all of which already have some form of protective status. According to Constitutional Article 129, "All activities with potential to generate damage to ecosystems must be previously accompanied by studies of environmental and socio-cultural impact (...)." In contracts that the Republic signs with natural or juridical persons, domestic or foreign, or in the permits it issues that involve natural resources, it is obliged to consider, even when not explicitly expressed, the obligation to maintain ecological equilibrium, to permit access to technology, and to return the environment to its natural state if it is altered, in terms set by law. But the Mining Arc directly contravenes such clear regulations. First, the decision to open nearly 112,000 km^2 to large-scale mining, and the convocation to international mining businesses, were all made before any kind of environmental impact study. According to the President of the Central Bank of Venezuela, Nelson Merentes, the government has already signed agreements with 150 national and transnational companies, "that, henceforth, will be allowed to begin exploratory work to certify mineral reserves, in order to then move to the phase of extraction of gold, diamonds, iron and coltan."[2] The specific content of these agreements is not public. According to Article 10 of the Decree, The Specific Development Plan of the Zone must be completed within a period of 6 months, starting with the publication of the pertinent decree in the Government Record (*Gaceta Oficial*). That Plan, including environmental impact evaluations, was supposed to be finalized in August 2016. If such extensive studies are supposed to be completed in such a short time, it is easy to imagine their superficiality and low quality.

Such a rush, though, facilitates the investment on which the decree depends. Encouraging and anticipating the participation of "private, state and mixed businesses," it offers a wide range of public incentives to mining corporations, including an easing of regulations; simplification and acceleration of administrative procedures; exemption to specified legal requirements in Venezuelan legislation; the generation of "preferential financing mechanisms;" and a special customs regime with tariff priorities for their imports. Article 21 also anticipates having a special tax regime with a total or partial exoneration of tax payments on rent and of the value-added tax. Within the framework of sectorial economic policy, the Federal Executive is authorized to grant total or partial tax exonerations on rent and the value-added tax, applicable to activities connected to mining, in order to stimulate the growth of the Orinoco Mining Arc National Strategic Development Zone. Such provisions enjoy a strong legal foundation, since the decree has the rank, value, and force of a founding law on an issue, known in most Latin American countries as organic law.

Likely Impacts of the Arc Project

Indigenous Peoples

The wide band of territory denominated as the Orinoco Mining Arc is inhabited by many indigenous groups, including the Warao, E´Ñepa, Hoti, Pumé, Mapoyo, Kariña, Piaroa, Pemón, Ye´kwana, Akawako, and Sanemá peoples. In addition to the potential devastation of their lands from planned mining activities, there would be increase in the scale of sociocultural phenomena that now accompanies mining activity everywhere: violence, corruption, prostitution, alcoholism, drug trafficking, rape of minors, paramilitary activity, and police and military repression. In short, such activity constitutes a flagrant violation of the indigenous peoples' rights guaranteed in Constitutional Chapter VIII. As Article 119 affirms, the state will recognize the existence of indigenous peoples and communities; their social, political, and economic organization; their culture, customs, languages, and religions; as well as their native habitat and rights over the ancestral lands that they inhabit and are necessary to guarantee their way of life. With the participation of indigenous peoples, the federal executive demarcates and guarantees the collective property rights to their lands, which are inalienable and non-transferable in accordance with the Constitution and the law. Article 120 states that the use of natural resources in indigenous habitats by the state will be conducted without harming their cultural, social, and economic integrity and only after informing and consulting the respective indigenous communities. In addition to these rights, equally violated are rights guaranteed in the principal legal instruments approved by the National Assembly (the country's legislative branch) in the 2001 Law *de demarcation and guarantee of the habitat and lands of indigenous peoples* and the 2005 Organic Law *of Indigenous people and Communities* (LOPCI). Among these violations special emphasis is on all the norms of prior consultation and information that are firmly established as much in Venezuela as international legislation (*Convenio 169* of the OIT) in cases of planned activities that may negatively impact the habitats of these peoples.

The National Executive did not consult affected indigenous peoples over the Orinoco Mining Arc decree, as Constitutional Article 120 requires, "in good faith" and "in the language" of each affected indigenous peoples, as detailed in Article 11 of Chapter II of the LOPCI. This egregious violation stems from, and technically justified by, its refusal to recognize them, as starkly illustrated by the fact that on December 15, 2016, it will be 15 years in which the National Executive will have failed to demarcate the lands and habitats of the indigenous peoples that inhabit territory in the Bolivarian Republic of Venezuela. The Twelfth Transitory Provision of the Constitution gives to the National Executive, for the Demarcation of Lands and Habitats of Indigenous peoples, a period of 2 years starting when it came into force on December 15, 1999. According to the Constitution, the Demarcation of Lands and Habitats of the 42 Indigenous peoples recognized in the Constitution should be finalized on December 15, 2001. The failure to do so, needless to say, violates the spirit and the letter of our Magna Carta. If the Federal Executive had honored the Constitutional and recognized indigenous peoples through the concrete and basic step of a Demarcation of their Lands and Habitats, the decree of the Mining Arc decree would not have been possible. The Federal Executive is willing to open millions of hectares of land to economic activity, but not to document the indigenous people living on them.

A relatively recent historical comparison helps clarify what this means. For the National government, headed by President Nicolás Maduro since 2013, the indigenous peoples whose ancestral homes are within the Orinoco Mining Arc decree zone seem to be the same type of "irrational peoples", as were the Warao People to the 1964-1969 government of President—Raúl Leoni. Without consulting the Warao communities, the government closed their access to the Caño Manamo, a branch of the Orinoco River in the delta area that is their long-standing home. This closure was decreed for the exclusive benefit of the Iron and Orinoco Mining Companies to enable them to remove, as much in the winter in the summer, the valuable iron minerals from de Bolívar Hill (Cerro Bolívar). The closure of the Caño Manamo generated massive desalinization that led to the loss of the Warao people's water and land, bringing with it death by hunger, thirst, and desperation.

If the very existence of indigenous people continues to be essentially ignored, it is now in the name of "21st Century Socialism," the reigning mantra of the government since the 1998 election of Hugo Chávez. But an equally profound casualty of the Orinoco Mining Arc decree may be the dream of a multiethnic and pluricultural nation that, according to the Preamble of the Constitution, we were going to be one day. The Orinoco Mining Arc decree places the country squarely into the rationality and culture of Western capitalism; those 112,000 km^2 contains gold, diamonds, coltan, uranium, iron, copper, bauxite, and rare earth minerals, and so the only rational approach is to extract and convert them into merchandise, even when this extraction destroys air, water, and life altogether.

Water, Energy, and Infrastructure

In the past decade, water has become an increasingly critical issue around the world as the increase in demand for water has risen along with the difficulty of accessing it. As climate change alters patterns of rain and drought, it reduces the availability of water for hundreds of millions of people, as a result of melting glaciers and the resulting decline in their capacity to hold fresh water. Human activities such as fracking and the massive use of toxic agricultural chemicals are reducing the volume of available water through irreversible levels of contamination. In the past, as much in Venezuela as the rest of the planet, priority was given to the exploitation of minerals and hydrocarbon in and underneath bodies of water, on the assumption that it was an infinitely available good. Policies based on this supposedly endless supply of water made by countries throughout the world have been catastrophic. The most dramatic example in Venezuela is Lake Maracaibo, from which the cities of Maracaibo, Cabimas, and other towns along its coasts get their water, but which, since the beginning of the twentieth century, have been sacrificed up ascollateral damage to oil extraction. In particular, the lake's huge Navigation Canal, constructed to enable access to large petroleum ships, has allowed the entrance of saline seawater. This mixing, together with agrochemical contamination and the discharge of untreated sewage waters has, over decades, slowly but surely been killing Latin America's largest lake. Is Venezuela society going to repeat such an environmental disaster in the Orinoco river basin? According to the Intergovernmental Panel on Climate change, Venezuela will experience increase in temperature and decrease in rainwater volume in the coming decades. The drought of recent years that lowered the water volume that gets to the Guri dam is thus a foreshadowing of what soon will be the new normal. The entire area of Venezuela, south of the Orinoco, is the country's principal source of fresh water, and the deforestation accompanying large-scale mining activity will inevitably and probably drastically reduce water volume in the zone.

Another nationwide impact from the decree is on electricity. One of the phenomena with the greatest impact on people in Venezuela in recent years has been the electrical crisis, due in part to the reduction of the flow of the Caroní, a river whose hydroelectric dams generate up to 70% of all electricity produced in the country. Along with the impacts of climate change, the large-scale mining in the Orinoco Mining Arc territory would severely reduce these dams' electric generation capacity. In the first place, as described above, would be the reduction in the flow of rivers in the zone impacted by the mining. In addition, mining activity above these dams, by reducing the trees and other forms of cover of the surrounding areas, would invariably increase the process of sedimentation of these dams. Together, all this would progressively reduce their storage capacity and useful life. All of the hydroelectric dams of the Caroni river system are located within the limits demarcated as part of the Orinoco Mining Arc.

Militarization of the Zone and the Prevalence of the "General" Over the "Individual Interest"

An additional threat to Venezuela's environment is the country's already-weak rule of law. Opportunities to oppose the negative impacts of big mining in the Mining Arc zone have been blocked by the decree's regulations. With the specific aim of forestalling any resistance to these activities, a Strategic Development Zone was created under the authority of the National Military (*Fuerza Armada Nacional Bolivariana*). According to this law, "the Bolivarian Armed Forces, together with the organized Poder Popular, and in coordination with the authorities of the Ministry of Popular Power with responsibility over petroleum policy will have the responsibility to safeguard, protect, and maintain the continued harmony of the operations and activities of the Strategic Industries located in the Orinoco Mining Arc National Strategic Development Zone, in conformity with that arranged in article 89 of decree N° 295 con Force and Standing of the Mining Law, Article 109 and others of the General Regulation of the Law of Mines, and article 40 of decree N° 1.395 with the Standing, Value, and Force of organic law which reserves the state the activities of Gold Exploration and Mining, as well as the Auxiliary and Related regulations."

Central to Venezuela's governing structure is the ideology of citizen power embodied by the 1999 Constitution. Adopted and approved following the election of Hugo Chávez, it added two branches to the government—the Citizen and Electoral branches—to the traditional executive, legislative, and judicial triad. Most innovatively, the Citizen's Branch is comprised of the Prosecutor General, the Defender of the People (ombudsman), and the Comptroller General, who together oversee adherence to the rule of law by governmental officials at all levels and so are empowered to prevent, investigate, and punish administrative irregularities. The extensive abuses allowed by the Orinoco decree, in the name of the Citizen Power that is supposed to prevent them, demonstrates how much this "power" has been coopted. For example, Article 25 of the decree expressly establishes the suspension of civil and political rights in the entire territory of the Mining Arc. According to this and other articles, no interests of individuals, unions, associations, are allowed to prevail over the general interest in the fulfillment of the decree's objectives. Individuals who carry out or promote conduct intended to obstruct all or part of the productive activities of the

Strategic Development Zone established in this decree will be penalized in accordance with the applicable judicial order. State security agencies are authorized to carry out and actions deemed necessary to safeguard the all activities foreseen in the Development Plans of the Zone of the National Strategic Orinoco Mining Arc, as well as the implementation of arrangements specific in this article.

The consequences of this "prevalence of the public interest over individual Interests" are extraordinarily serious. "General interests" are understood as mining as conceived in this presidential decree. Every other vision or interest, including appeals to the Constitution, are defined as "individual interests," and thus subject to the "organs of State security" carrying out "immediate actions necessary to safeguard the normal implementation of activities specified" in the decree. But what are or might be the interests labeled as "individual?" The decree is written in a way that allows for a wide interpretation. For example, it expressly defines union interests as "individual." This view can, without doubt, justify the suspension, in the entire zone, of labor rights guaranteed by Constitutional Article 43 and the *Organic Law of Work and Laborers*. Does this equally imply that the "union" rights, and thus the "individual" rights of journalists to report from the zone also remain suspended?

What implications does this have for the indigenous peoples, who would undoubtedly be the population most affected by these activities? Would the constitutional rights of those people, following state "regulations" understood as "individual interests," have to be repressed if they conflict with the "general interest" of mining in their ancestral lands? The answers to this question are even more worrying in light of the fact that in February 2016, just 2 weeks before the decree establishing the Mining Arc Development Zone, President Nicolás Maduro decreed formation of what it calls the Anonymous Military Company of Mining, Oil, and Gas Industries: (Camimpeg: *Compañía Anónima Militar de Industrias Mineras, Petrolíferas and de Gas*), under the Ministries of the Citizen Power and of Defense. This Company has a wide spectrum of authorities to concentrate "without any limitation" on any activities related directly or indirectly with mining, oil, or gas.[3] With the planned participation of this business in the activities of the Mining Arc, the armed forces, far from representing the defense of a hypothetical "general interest" in the zone, will have a direct economic interest in which mining activities would face no kind of obstacle. They would, according to this decree, be legally authorized to consequently act as they see proper in defense of those interests.

In fact, through presidential decree, Venezuela has found itself with the suspension of the 1999 Constitution in 12% of its territory. While citizen rights will be truncated, the government can pursue at least two objectives. First, it will grant guarantees to transnational businesses whose investment they are looking to attract, enabling them to operate freely without risk of facing resistance to their activities. Second, it gives the military even more power within Venezuela's state structure. On both regards, built into the decree is the criminalization of the anti-mining resistance being mounted by rural workers, indigenous peoples, popular movements, and organizations throughout the continent.

Conclusion

In conclusion, a government that calls itself "revolutionary" and anti-capitalist has decreed the country's subordination to the interests of major transnational mining corporations with extractivist predatory activities that jeopardize the future of the country and put its indigenous peoples at risk of ethnocide. All these reasons make it urgent to demand a complete derogation of the Orinoco project, or submitt it to public vote through a consultative referendum. Venezuela's environmental policy must instead be oriented toward monitoring and control, as well as formation of market incentives that reign in excessive use of resources and of ecologically damaging production techniques. Such incentives would also support the incorporation of environmental protection into the private sector and help finance the State's own protective work. At the difficult crossroads where it now finds itself, Venezuela must invest time, money, and resources into environmental protection. It can do so externally, such as by providing funds for countries that attain certain environmental goals, as well as internally, such as promoting Venezuela as a center of ecotourism and ecological research.

Through its description and analysis, this chapter initiated a constructive debate intended to reconcile positions and interests geared toward the conservation and improvement of the quality of life of our country's citizens and the defense of our Amazonia, as well as to explore mechanisms and actions that can be undertaken by both state officials and environmental groups. In particular, such agencies should propose an international regulatory framework to legislate sustainable exploitation of strategic resources, some of which are not renewable. Given the Venezuelan government's current recklessness, only such collective oversight can preserve the Amazon.

Notes

1. *Agencia Venezolana de Noticias*, "Gobierno nacional prevé certificar en año and medio reservas del Arco Minero Orinoco," Caracas, February 25, 2016.
2. *Agencia Venezolana de Noticias*, "Plan del Arco del Orinoco contempla industrializar potencial minero nacional," Caracas, 27 de febrero de 2016.
3. This list of activities is extraordinarily exhaustive, connected to any activity—investments, marketing, exports, and financing—connected to these industries.

CHAPTER 8

Suriname: An Exposed Interior

Katia Delvoye, Minu Parahoe and Hermes Libretto

Abstract Examining the Amazon's often overlooked non-Latin corner, this chapter first describes the economic, social, political, and environmental context of the country of Suriname and its Amazon rainforest interior. It then lays out the country's environmental legislation and regulations, and the effectiveness of their enforcement, particularly by indigenous and other authorities. In their assessment of this legal and policy structure, the authors examine how rainforest conservation organizations handle the many conflicts, opportunities, and challenges they face.

Keywords Amazon · Environment · Indigenous peoples · Rainforests Suriname

Introduction: Political, Geographic, Socioeconomic, and Environmental Contexts

The Republic of Suriname is located on the northeastern South American coast, bordered by French Guiana, Brazil, Guyana, and the Atlantic Ocean. The country has four main ecological zones. The northern young coastal zone is characterized by mud flats and sandy strips

K. Delvoye (✉) · M. Parahoe · H. Libretto
Amazon Conservation Team, Paramaribo, Suriname

M. Ungar (ed.), *The 21st Century Fight for the Amazon*,
https://doi.org/10.1007/978-3-319-56552-1_8

Fig. 8.1 Map of South America with Suriname in *Green*; map of Suriname

comprised primarily of woodland vegetation and mangrove forests. Further inland lies the old coastal plain and the primarily white-sand (bleached) savanna belt. The remaining southern zone, covering 80% of national territory, is covered with dense and pristine tropical rainforest with mountain ranges and numerous waterways that mostly flow in a south–north direction, discharging approximately 5000 cubic meters of fresh water into the Atlantic Ocean per second. With a mean annual air temperature of 81° F (27° C), Suriname's climate is a humid tropical type, influenced by the Inter-tropical Convergence Zone, with two rainy and two dry seasons annually over most of the country. Suriname has a rich biodiversity and endemism of flora and fauna. Over 5800 species of mosses, ferns, and seeds plants are found in the country, of which an estimated 50% are endemic to the Guiana Shield region (Alonso and Mol 2007), a section of the South American tectonic plate comprised of Suriname, Guyana, French Guiana, and the Brazil's Amapá state. Of the nearly 2000 known vertebrate species, 3% are specific just to Suriname (Fig. 8.1).

This dense concentration of rich biodiversity, though, faces many threats. Approximately 13% of Suriname's land area is under some form of legal nature protection. But its southern rainforests, like those across the Amazon Basin, are being impacted by multiple forms of destruction in the Guiana Shield, primarily deforestation and forest degradation caused by industrial and illegal small-scale mining. This pressure is considered to be the fastest growing cause of forest loss in the Guianas. Above all is gold mining, which doubled in Suriname between 2008 and 2014, with nearly 99% of it in the Greenstone Belt, a geological formation known for its large gold reserves. Such heavy extraction, and the logging that accompanies it, results in degradation of the hydrology, turbidity, and nutrients of streams; introduces heavy metals; and harms aquatic vegetation and animal life. In particular, small-scale gold extraction in Suriname, which is almost always illegal, uses mercury, which is highly toxic and stays in the ecosystem for disconcertingly long periods of time (Ouboter 2012). In Suriname's southern region, the livelihoods of indigenous populations and Maroon people (tribal groups formed by the descendants of escaped African slaves), who are concentrated along inland waterways, are most directly affected by mercury pollution and accumulation in the food chain.

Two laws, the Labor Law and the Mining Decree, include regulations to manage mercury's use and reduce its adverse health effects. But the country's legal structure is not yet able to implement such regulation or otherwise reign in such activity. While the import of mercury requires a license, for example, no additional law explicitly regulates or even mentions it. Within the national government, more broadly, responsibility for and knowledge regarding mercury is fragmented among ministries and departments, which hampers policy control over its use (see Heemskerk et al. 2015). Such oversight is fragmented vertically as well as horizontally. The 1987 Constitution, amended in 1992, established Suriname as a democracy, with a government led by a president and a 51-member parliament called the National Assembly. The President appoints and oversees the Council of Ministers, ratifies approved bills, proposes State decrees. But day-to-day governance is increasingly decentralized, mainly through a process that began in 1987, centered on three levels of government: national, district, and sub-district. District government is split into District Administration, which is its executive branch, and the District Council, the legislative branch, responsible for district-area laws and regulations. The sub-district level has only a representative branch, the Resort Council, whose main responsibility is to advise

District Administrations and Councils. There are ten district councils and 62 Resort Councils. Although this structure has been adopted, financial decentralization has been only partially implemented, seriously undercutting this new district governance.

Suriname's economy is dominated by the mining industry, with gold and oil accounting for about 85% of exports and 25% of government revenues, making the economy highly vulnerable to mineral price volatility. Fisheries, tourism, agriculture, and forestry account for less than 10% of GDP. The country's tourism industry centers on a network of protected areas and river resorts, particularly in the forested interior, while the coastal zone is also an international wildlife destination with outstanding birding potential. The agricultural sector employs approximately 25% of the population, and is concentrated almost exclusively within the coastal zone; the predominant form of agriculture in the interior, in contrast, is subsistence cultivation of rain-dependent crops. The government sector employs 40% of the labor force, while the 2015 unemployment rate was approximately 9%. Amid these limited labor opportunities, remittances from the many Surinamese working abroad are critical, bringing an estimated US$140 million into the economy each year. During the global economic crisis that erupted in 2008, though, Suriname maintained macroeconomic stability despite a recent history of double-digit inflation, volatile investment ratios, and a weak macroeconomic policy framework. That growth was underpinned almost entirely by increasing production of minerals with high prices. But due to declining oil and gold prices and high state expenditures since then, the economy is now facing monetary pressures after suffering a currency devaluation of 124% in 2015.

Suriname's mineral wealth is seen in the extensive mining of oil, gold, bauxite, gravel, sand, kaolin, and other minerals. The three biggest of those commodities were gold, oil, and alumina until September 2015, when the Surinamese bauxite mining company Suralco, part of the multinational firm Alcoa, halted its mining of bauxite (a mixture of aluminum oxides and other minerals) in the country, leaving oil and gold as the economy's dominant activities. Altogether, such extraction accounts for more than 50% of GDP, over 90% of total exports, and between 30 and 35% of government revenues. The biggest growth area is oil production in Suriname's coastal zone. The state oil company, Staatsolie, has exclusive rights to this zone, with most exploration, extraction, and production taking place in its wetlands. Staatsolie's first oilfield came on line in 1982, and, according to company reports, its crude production totaled 5.9 million barrels in 2015 with an average daily production of 16,000 barrels. The company initiated a diversification program in 2011 and has stated that between 2015 and 2020, it will invest over 1 billion US dollars in expansion of its operations. It also plans to further increase its oil reserves through nearshore drilling and expansion and consolidation of oil production. It will also grow through purchase of a 25% stake in the ambitious Merian Gold project for extraction in a zone with an estimated 5.1 million ounces of gold reserves. The government participates with a 25% equity ownership stake in the Merian open-pit gold project, through Staatsolie Maatschappij Suriname N.V., owned by the state of Suriname. Canadian Newmont will be the managing partner, with a 75% interest. Staatsolie has made an initial cash contribution of approximately $83 million to Newmont, which estimates that total capital investment for the Merian project will be between $900 million and $1 billion (E & MJ Engineering & Mining Journal, http://www.e-mj.com). But, as in other countries, this project could quickly incite conflict and damage, so these companies will need to maintain transparent communication with the local communities, which have long been engaged in artisanal mining and are wary of large-scale projects.

The great majority of Suriname's 542,000 people live in the 30-km-wide Atlantic coastal zone, in and around the capital city of Paramaribo. South of the coast, population density declines abruptly. The southernmost district, Sipaliwini, which covers nearly a quarter of the total national area, is home to just 6.8% of the population. Society is ethnically diverse, consisting of Hindustani, Creoles, Javanese, Maroons, indigenous peoples, smaller groups of Chinese, Lebanese, Europeans, and people of mixed descent. Amerindian and Maroon communities live mainly in the southern rainforest. Since the 1960s, but particularly since the mid-1990s, others have come to work and live in that remote area. They include Brazilian small-scale gold miners, Chinese store owners and loggers, foreign missionaries, development workers, nurses, and teachers from the city. As in other countries in the Amazon, as discussed in this book, they seek economic opportunity. On the 2016 United Nations Development Index—a composite statistic of life expectancy, education, and per capita income indicators—Suriname ranks 97 out of 187 countries. But this internal migration will do little to push Suriname up in the ranking. Levels of literacy in the adult population is approximately

90%, but in the southern interior, educational levels and services are far below national standards. Although gold mining contributes to economic development in the region, particularly through formal and informal employment, surveys among the southern tribal communities reveal that mining barely compensates for its damages on the forest ecosystem and for its many negative social consequences, such as alarming increases in crime, smuggling, violence, and job-related accidents. Around the region, water pollution and disturbance of waterways has led to health problems such as diarrhea and malaria, while mercury poisoning contributes to illnesses like chemical bronchiolitis and pneumonitis (from inhalation) as well as damage to the lungs, kidneys, the nervous, digestive, and immune systems.

Environmental Legislation, Regulations, and Enforcement

Environmental legislation in Suriname is still fragmented into many laws and regulations. The Nature Preservation Act of 1954 establishes arrangements for the protection and maintenance of natural monuments; the 1973 National Planning Act sets out national and regional planning on subjects like land use policy; the 1992 Forest Management Act created a framework for forest management and sustainable use of forest resources; and the 1998 Governmental Decree on Nature Protection established the Central Suriname Nature Reserve. Regulating resource extraction are the 1986 Mining Decree (dealing with bauxite, gold, sand, gravel, and water), the 1954 Game Act (updated in 1997), and the 1961 Act on Fisheries (reformed in 1981). Some forest- and mineral-based Acts were updated in the late 1990s. In an attempt to achieve coordination between the different ministries dealing with this wide range of issues—reflecting the fragmentation discussed throughout this book—inter-ministerial commissions were formed between 1990 and 1998, along with the National Environmental Board (NMR). But because of the slow work and few results of the commissions as well as the NMR, the National Institute for Environmental and Development in Suriname (NIMOS) was established in 1998 to initiate a legal and institutional framework for environmental policy with regard to sustainable development. To coordinate and update all these laws, in addition, a legislative process that began in the early 2000s led to a draft Framework Act on Environmental Management along with draft regulations on pollution control and environmental impact assessments. But to date they remain to be discussed and approved by the National Assembly.

Undelying the impetus for reform is the fact that current legislation reflects perceptions of the pre-colonial era and insufficiently addresses environmental prevention, pollution, and mitigation. But while the Environmental Framework Act remains mired in discussion, mainly because it is not a priority for the government, NIMOS did develop and implement a set of guidelines for environmental impact assessments for different categories of projects. For multinational development projects, in particular, these guidelines require firms to comply with international standards, and, to get and remain certified, their operations must strictly follow environmental and socially responsible rules. But for implementation of major projects, the inclusion of environmental provisions in project preparation, execution, monitoring, and evaluation remains voluntary. The lack of clarity resulting from this fragmented legislation and regulation, as well as from poor inter-institutional coordination, leads to many environmental violations and conflicts between tribal communities and entrepreneurs. The broadest and most ambitious law for nature conservation, sustainable development and sustainable resource is the Environmental Framework Act, drafted by NIMOS in 2002, to regulate pollution, waste management, and environmental impact, among other subjects.

Environmental policies fall under the Cabinet of the President, which authorizes NIMOS to draw up environmental legislation and to coordinate its implementation. NIMOS also reviews and monitors Environmental (and Social) Impact Assessments (E[S]IAs), and gathers and disseminates data on land use and management. Different ministries and semi-governmental organizations, such as the Foundation of Forest Management and Production Control (SBB) and the Suriname Foundation for Nature Management (Stinasu), oversee specific environmental or nature management mandates. The Ministry of Physical Planning, Land and Forest Management (ROGB) formulates national policy on land use planning, as well as sustainable forest use and nature conservation, with subdivisions for regulation, implementation, monitoring, and control. The Ministry of Finance, along with the National Planning Office (SPS), prepares development programs for the national government; has a unit for Environment and Spatial Planning to incorporate the environment into physical planning; maintains data on land and soil, natural resources, existing infrastructure, and land allocation;

and maintains geographical data in order to map ecosystems and demographic and physical indicators. The Ministry of Natural Resources (NH) manages the development of water and regulates domestic, public, and commercial energy use. The Ministry of Agriculture, Animal Husbandry and Fisheries (LVV) regulates the agrarian production sector and the proper use of agricultural lands and waters. The Ministry of Public Works (OW) plans and implements civil engineering and infrastructural works, water management and drainage, hydrological and meteorological services, and waste management. At the district level, District Commissioners (DC) under the Ministry of Regional Development (RO) are the main representatives of the government and work closely with the elected District Council to coordinate general management, economic promotion, and development of the district. Together, this wide range of national and local institutes aggravates the lack of clarity over their overlapping responsibilities. Coordination among them is consequently weak and often leads to misunderstanding, inefficiency, and conflict. Many clashes erupt, for example, when concessions are issued in buffer zones of special forests.

As in other countries, one of the main strategies to minimize such dysfunction is creation of protected areas. Suriname has created three types: Nature Parks, Nature Reserves, and Multiple Use Management Areas (MUMAs). Nature Reserves are locations with significant biodiversity and/or geological, archeological, or cultural attributes. They are managed as high-value natural areas with fairly restricted use. For instance, the Nature Preservation Law (1954) forbids persons "to either deliberately, or through negligence, damage the soil conditions, the natural beauty, the flora and fauna, or to perform any action which destroys the value of the reserve." Hunting, fishing, camping, and some other recreational activities may be conducted only with written permission from the Forest Service. Nature Parks are relatively low-level conservation areas. Suriname has only one Nature Park (Brownsberg Nature Park), managed by STINASU, a government foundation for nature preservation. Multiple Use Management Areas are designated to maintain biological productivity, ensure the health of globally significant wildlife, and protect resources for sustainable livelihoods. Despite this careful designation, though, most of Suriname's protected areas are "paper" parks, since they lack any real management presence, with a perennial lack of personnel, vehicles, and reliable equipment.

Enforcing all these environmental regulations (as well as all other laws) is Suriname's Ministry of Police and Justice. Within that ministry, the many subdivisions of the national police department handle different responsibilities. The City Police covers the capital city; the Rural Police manages issues beyond the capital; and the Judicial Police is composed of specialized units, such as fraud and homicide. The second largest law enforcement body is the Military Police (MP), a unit of the Surinamese Armed Forces, which oversees all members of the Surinamese military and handles border control and immigration functions—a critical role, given the levels of smuggling around the region discussed in this book. The third largest enforcement body is the Surinamese Agency for National Security, responsible for the Central Intelligence and Security Service (CIVD).

As part of decentralization, most policing occurs at the local level. Most significantly for the environment, police stations in all ten of the country's districts serve as potential control points. Game wardens of the Forest Service (LBB), for example, have posts in many protected areas and along public roads. These wardens may act against violations of the Game Act, the hunting calendar, the Nature Conservation Act, and the Fisheries Act (negotiated with the Ministry of Agriculture), as well as against any international conventions and agreements to which Suriname is a party, notably CITES (Convention on International Trade in Endangered Species) and Ramsar (The Wetlands Convention), which also plays a role in Ecuador's environmental enforcement, as Chap. 5 discusses. Map 1, below, shows the sites where game wardens are stationed in order to regulate, control, and inform communities and private businesses. The map also distinguishes the northern part of the country to which the Game Act applies, and where the coastal zone is relatively better prepared for identifying, researching, and violating environmental offenses. But in the adjacent landscapes, particularly in the savannas and forested interior, the presence of enforcing bodies is low to non-existent. A primary reason is the difficulty of Suriname's government to fund the expensive maintenance of field staff in the southern region. In fact, the high logistical costs for travel, lodging, and on-the-ground operations for those regions exceed the limited budget for the country's entire nature and environment legislation enforcement effort (Fig. 8.2).

To monitor and control the timber industry, Suriname's National Foundation for Forest Management and Production Control (SBB) has several posts along routes connecting forests and sawmills. They

Fig. 8.2 Game wardens Station Locations. *Source* Centre for Agricultural Research in Suriname (CELOS)

also conduct regular controls on timber concessions. The Gold Mining Reform Commission (OGS) is charged with reforming the gold mining sector under the Ministry of Natural Resources, and works with police and military units to man remote posts in the mining areas. The Geological Mining Division (GMD) also has an inspection division to enforce compliance with mining regulations. Even with these agencies, though, Suriname's government still lacks a broad environmental policy, and no environmental police units have been developed to improve environmental prosecution. The other set of entities that help enforce environmental law are NGOs and other efforts within society. Although environmental awareness and societal engagement have both grown considerably in recent decades, civil society involvement remains limited. Community-based resource monitoring is only marginally developed. In the coastal zone, the Kaliña indigenous people collaborate with the Forest Service (LBB) and WWF Guianas on sea turtle monitoring. The Green Heritage Fund Suriname (GHFS) works with volunteers to monitor dolphins and sloths, and cooperates with the LBB and fisheries bacnch of the Ministry of Agricultures to report breaches observed in the marine/estuarine zone. In Suriname's southern region, the Amazon Conservation Team (ACT) sponsors a unit of 27 rangers or "indigenous park guards" (IPGs) of the Trio and Wayana indigenous peoples. These IPGs have no formal mandate recognized by the LBB, but can report abuses of wildlife trade law and conventions to authorities. The IPGs have also assisted the State Commissions in their regulation of the gold mining sector. A collective 2014 initiative led to a reinforced proposal, the South Suriname Conservation Corridor (SSCC). By linking the SSCC to the Central Suriname Nature Reserve (CSNR), this project will ensure the protection of two-thirds of Suriname's headwaters and some of the most pristine rivers on Earth.

Environmental Enforcement in the Interior and the Role of Traditional Authorities

National Decree C17 of 1983 delineated the boundaries of Suriname's ten administrative districts. The interior of Suriname is designated in two ways (Healy, Libretto, 2010): by either the cartographically southern parts of the country (south of 5° latitude) or the administrative area in which tribal communities live, which is mainly the interior but also includes parts of the coastal plain where Maroon and indigenous communities have settled and continue to use traditional political systems. To help implement the decentralization plans launched in the 1980s, the government established a Ministry for Interior Development in 1989 and provided greater recognition of tribal chieftaincy. Thereafter, two leadership systems existed at the local level: a formally recognized decentralized government administration, and a traditional authority system not acknowledged in legislation. Amid insufficient engagement and dialogue, tension quickly arose between these two systems. Unclear decisionmaking on the district level, in particular, has triggered conflicts when permits and concessions are issued in Paramaribo without the knowledge or consent of regional administrations, let alone of traditional authorities. In addition to the limited educational and economic opportunities available to their communities, local authorities express concerns regarding potential environmental damages, disruption of social structures, and weak enforcement of existing legislation to protect local livelihoods.

In general, Surinamese legislation applies to the entire territory. The only exception is the Game Act, for which the State distinguishes a northern and southern zone in which the Act applies. In the southern zone, the Act allows tribal communities—who have poor access to markets and practice subsistence livelihoods—are allowed to hunt for protein sources around their villages. But if anyone in possession of seasonally or permanently protected species crosses from the southern zone to the northern zone, they are in violation of the law. The closest police station and game warden surveillance squad can enforce these regulations. All environmental legislation besides the Game Act applies to the nation's interior, and is supposed to be enforced by the respective environmental agencies. The Mining Act, mentioned above, governs current and potential economic activities encroaching on the interior communities. Permits for such activities are to be issued only through the Ministry of Natural Resources and its Geological Mining Division (NH/GMD) in Paramaribo, and only after consultations with the district administration, potentially impacted communities, traditional authorities, and other stakeholders. Most such consultation is insufficiently carried out in practice, however, leading to situations in which entire villages are included in the approved activity zones of issued permits without their approval. For industrial-scale operations, NIMOS requires environmental and social impact analyses (ESIAs). In addition to mining, NIMOS also requires ESIAs for logging operations, infrastructural development, and hydroelectric power plant development.

The highest authority in Suriname's Maroon and indigenous ethnic groups is the *granman*, the paramount chieftain, who is assisted by a council of elders, *head-captains*, and *captains* (village chiefs). The *granman* and *captains* are assisted by *basias*, who are administrative assistants. Traditional authorities are elected by their tribal group, and hold those positions for life. The leadership institutions of the coastal tribes, in contrast, are much less hierarchal than that of their southern counterparts. In most of those areas, individual tribal chieftains must deal with ever-changing challenges without the support of advisors and legal experts. In particular, chieftains must combine the governance roles associated with traditional indigenous and Maroon culture with policy development and relationships with the national government and international businesses.

The duties of Suriname's traditional authorities can be divided into the following four roles:

1. Political and administrative: This role centers on maintaining relations with the national government. In Suriname, the relationship between the traditional authorities and the national government is based largely on custom and oral agreements, since principles and procedures are not defined in the Constitutions or national

laws. Traditional authorities receive an honorarium from the government, but functional relations between the village chiefs and national officials are not structured or evaluated. Within this undefined context, the day-to-day work of the traditional authorities is greatly undermined and often paralyzed by their lack of administrative resources and operational budgets. Daily governance, handled at the village level, is also affected. Decision-making may require days of gathering with a range of groups, from specific families to entire villages.

2. Sociocultural: Traditional authorities take care of the well-being of their tribal communities and protect them from adverse external influences. Traditionally, this role included redistribution of resources to the weakest in the community, and dealing with governmental and non-governmental institutions promoting social development. Unfortunately, though, some Surinamese traditional authorities disregard these many responsibilities and instead use their privileged position for the betterment of their kin group, including promotion of family members for jobs in the government or NGOs. These authorities are often the religious leaders of their group, in addition, leading to further bias against other religous groups.
3. Juridical: Under current legislation, traditional authorities have no mandate to dispense law and justice, but Indigenous and Maroon peoples do have a right to apply their own legal and juridical systems within their territories, as long as they are in accordance with national law. Local authorities also play a crucial role in maintaining law and order, because for many villages, the nearest police post is many hours if not days of travel away. As political and social leaders, traditional authorities are generally seen by their communities as having the additional responsibilities to ensure sustainable use of natural resources, to establish and endorse rules that support environmentally sound practices, and to sanction those who violate these unwritten rules. National state officials also allow these local leaders to resolve offenses such as physical assaults, thefts, or violations of customary land use rules, using a limited set punishment after consultations in village meetings. But these leaders do not have a formal mandate to arrest offenders or impose penal law sanctions, which severely limits their options to maintain order and justice, especially when involving non-local parties that neither

recognize nor respect the customary laws. Examples are the growing intrusion of illegal artisanal and small-scale gold miners and the huge impact of drug use. Traditional authorities are skilled in dealing with longstanding issues, but may not be as effective in handling those more recent problems. State law enforcement agents are only called in when a serious crime occurs, when a criminal act involves one or more outsiders, or when the traditional authorities feel that they do not have the capacity or resources to deal with the problem. In those cases, national police (or military) involvement is typically combined with customary responses.

4. Land stewards and managers: For centuries, tribes throughout Suriname have continually created and adapted land and resource utilization practices to ensure physical and cultural survival. Examples include diversification of hunting practices, fishing skills, use of common land. Adapation of a wide range of customary practices to regulate natural resources, in particular, shows that traditional authorities could take leadership in a sustainable land use system.

Much of the current framework was created in the national constitution and the 1992 Peace Agreement between the Government and the Jungle Commando, which fought to ensure equal rights for the Maroon peoples, along with the Amerindian armed organized called Tucayana Amazones and other groups, to put an end to the war that waged in the interior since 1986. The agreement provided a framework for the juridical role of traditional authorities, including resolving land rights issues (article 10). According to the new rules, all land in the interior is owned by the State or contracted as a concession or permit. Although tribal peoples do not have a formal right to utilize the lands they occupy, the government can designate economic zones where the communities of the tribal citizens would be able to engage in economic activities like forestry, small-scale mining, fisheries, and hunting. Suriname's traditional authorities oversee management of "community forests" (*gemeenschapsbos*), which the 1992 Forest Management Act mandates are to be allocated and sustained for the well-being of the entire village. Given their location, tribal communities of course rely heavily on natural resources in the forests for subsistence, whether in terms of cultivation plots, non-timber forest products (NTFPs), or protein resources. Because there are no formal regulations regarding their agreements with non-local parties who seek to use such forests commercially (e.g., timber harvesting), tribal chieftains can arrange with such interests to pay compensation to the village. In practice, though, a lack of regulation over such funds restricts their access by, and benefits to, the overall community.

Suriname's tribal chieftains are also responsible for relations with both large- and small-scale mining enterprises in the area they inhabit. Most small-scale mining operations in Suriname work informally and without legal documentation, with virtually no commitment to national regulations or supervisory inspection by national authorities (Heemskerk 2012). With respect to logging, in some past cases, tribal chieftains have granted permission to outsiders without the knowledge of community members. This also happens with permission granted for mining (gold, diamonds, and bauxite) in the surrounding areas. Similar gaps between law and practice characterize protected land. Of Suriname's 16 existing and four proposed protected areas, 15 are located within or near indigenous and Maroon peoples' traditional lands. These peoples have not always participated in decision-making processes regarding establishment of protected zones, and still do not participate in current management systems. The Nature Conservation Act of 1954 (see 2.1), which forms the basis for the establishment of National Parks and nature reserves, is thus felt to be culturally insensitive by many of Suriname's tribal communities. Most of the reserves that were declared as protected areas in the 1970s, based on 1950s legislation, did not involve much community engagement. Of particular concern to tribal area inhabitants are the restrictions placed on traditional land and resource use. For example, the community was no longer permitted to hunt and commercially fish within the boundaries of the protected areas.

Suriname's traditional tribes still apply their own forest rules and hunting calendar, applying unwritten rules to prevent biological resources from being overharvested. They reinforce standards that the killing of young and female (pregnant) animals should not be killed; that community members may remove from the forest only what they need and will use; and that for each game animal, there are well-defined and limited hunting seasons, with some areas completely off limits (Colding and Folke 2001). Surveys also reveal that the indigenous peoples of the southern region are using traditional knowledge to adapt to changes in rainfall and an increase in floods induced by climate change by revising requirements for securing food and livelihoods. For example, adjustment strategies adopted by Trio indigenous community of the southern region, in response to documented alterations in the life cycles of fauna and flora as well as a higher occurrence of crop diseases and soil degradation, include relocation of cultivation plots to higher elevations, crop diversification, and adjustment of the planting season.

As in other countries, though, Suriname's enforcement structure does not keep up with such practices. For example, the Game Act regulates hunting of wildlife only in the northern zone, providing the interior communities room for traditional harvesting of species for subsistence. With the absence of official enforcement authorities and weak information sharing, though, the desire for income generation by some members of the tribal communities has increasingly led to commercialization of bush meat and wildlife trade. Population studies, species distribution surveys and related assessments are not regularly conducted, which together deprive tribal communities with information on acceptable quotas per species and required criteria for entering the commercial sector. Only with such information—along with viable alternative income sources and strict fines for violations—can the communities fully protect the environment.

Mining licenses issued in or close to the land used by Suriname's tribal communities, in addition, have major environmental and social impacts. In particular, weak regulation of the extraction sector has led to large-scale degradation, deforestation, and pollution. Conflicts arise in tribal communities because of weak enforcement of license regulation, perceived inequities in benefit-sharing, and especially the confusing overlap of timber harvesting and mining licenses—not to mention the destructive impacts of the uncontrolled use of mercury and cyanide. Additional inequality and damage arise from management of the *gemeenschapsbos,* which, as mentioned above, refer to tribal societies' collective rights over use of forest resources, which is put into practice in part through permissions to pursue activities of sustainable development. But these permissions are often sold or transferred to third parties with financial capital, which then minimizes benefits to local communities while inflaming social and environmental disruption. The government has little control over such transactions, since the community forest is monitored solely by the SBB (Suriname's National Foundation for Forest Management and Production Control) and only over the commercial harvesting of timber resources. The tourism boom in the interior, concentrated in river resorts along the upper Suriname River, raises parallel issues. In the continuing absence of strict regulation of the tourism industry and enforcement of environmental laws, problems have grown over waste management, cultural identity, and the integrity of biodiversity. Equitable benefit sharing with respect to commercial and community-based tourism, moreover, has yet to be assessed or developed in any meaningful way. To date, just a few tourism enterprises operating in the interior have established a baseline for long-term community development. Even when national state agencies are able to reach the tribal communities to raise awareness and provide guidance, language barriers and cultural differences may negatively impact the results.

In the face of such challenges, though, are many emerging examples of effective leadership practices by Suriname's tribal communities to limit unsustainable activities. In eastern Suriname, several village councils have joined forces against illegal Brazilian gold miners, and in south-central Suriname, the indigenous *granman* (chieftain) Nowahé has rejected all requests from gold miners to work in the Wayana area. There are also cases in which state law enforcement agents have been summoned when traditional authorities feel that they do not have the capacity or resources to deal with a particular problem on their own. For example, *granman* Asongo of the Trio tribe has called for police assistance to investigate the production of, trade in, and use of marijuana, an increasing problem in the village of Kwamalasamutu. Throughout the southern region, police are also increasingly needed to deal with violent crime within the (gold) mining sites. In many of these cases, Suriname's tribal communities have been overpowered by other parties with far greater power. Summarizing the limits of traditional chiefs, Healy and Libretto (2010) write that they "have become development workers," even though "their customary training has not prepared them for this task. To date, they are not prepared for the new elements such as report writing, processing expenses, and international standards."

Rainforest Conservation Organizations and Enforcement of Environmental Legislation

Surinamese civil society organizations involved in conservation can be categorized by their target groups, geographic scope, and operational size. There are only about ten such organizations active in the rainforest regions, where they advocate wise use policy, sustainable resource extraction, and monitoring of indicator species. Several organizations carry out environmental education and outreach on a national scale without limiting themselves to a specific sub-region. They also help governmental agencies raise awareness regarding renewable energy sources, banning mercury use, waste management, recycling, sustainable use of non-timber forest products (NTFPs), and NTFP marketing. On the local level, though, only about five community-based organizations (CBOs) are structurally engaged in direct conservation activities—another indication of the gap between national and local power.

Many international organizations work to shore up these efforts. They include the World Wildlife Fund (WWF) Guianas, Conservation International, Tropenbos International and the ACT, with offices in the capital and work projects and partnerships around the country. The UNDP and the Inter-American Development Bank (IADB) broadly support policy development, such as facilitating dialogue on environmental issues and creating long-term mechanisms for environmental hazards, community resilience, climate change adaptation, and biodiversity conservation in the protected areas system. In 2000, The Global Environmental Facility (GEF) pledged approximately US$15 million primarily for management of the protected areas. With financial support from the UNDP, the Dutch Embassy, and the Suriname government, this endowment was placed under the Suriname Conservation Foundation with a focus on the Central Suriname Nature Reserve (CSNR, declared in 1998) and the Sipaliwini Nature Reserve (SNR). The fund has a mandate to support efforts by the Forest Service (LBB), as the central authority in protected areas, to manage them responsibly, in part through enforcement of legislation like the Nature Conservation Act. The fund is required to work within the framework of Suriname's 1954 Nature Conservation Act, which declares that the management of Suriname's protected areas is a State responsibility. A more recent player in this field is the Suriname Environmental and Mining Foundation (SEMIF), established in 2008 to develop the mining sector. SEMIF has since evolved into a fund that emphasizes sustainable development with the involvement of Ministry of Natural Resources as well as with both State and multinational mining companies. As part of that effort, this foundation also makes grants to efforts of tribal communities and NGOs related to biodiversity conservation.

Environmental legislation has not evolved much, meanwhile, but community engagement has filled the vacuum. Increasingly, organizations that address conservation in Suriname have become central through their promotion of sustainable livelihoods, property rights, and community empowerment. Prominent organizations within this category include the Women's Movement for Entrepreneurship (NVB), the Association of Indigenous Village Leaders in Suriname (VIDS), and the NGO Platform (the Bureau Forum NGOs). Some of their most effective work is economic. For example, the NVB and the NGO Platform have developed many (non-timber forest) product chains in the Samaaka Maroon communities. These organizations have made legal advances. One example was when VIDS submitted and won the case *Kaliña and Lokono Indigenous Peoples versus Suriname* at the Inter-American Court of Human Rights. Asserting that basic rights were violated, VIDS convinced the court that the State of Suriname failed to recognize and guarantee the legal personality and territorial rights of the Kaliña and Lokono, violating those and other rights in connection with bauxite mining, grants of individual titles to non-indigenous persons, and both the existence of and restrictions imposed in two nature reserves (Forest Peoples Program 2016).

But difficulties still far outnumber these few victories. Surinamese NGOs surveyed by the ACT in 2015 reported many challenges to environmental enforcement in the interior, ranging from weak inter-organizational collaboration to language barriers isolating Maroon and indigenous societies. In particular, NGOs say that government agencies focus on timber extraction without fully acknowledging or incorporating the legal status of *gemeenschapsbos*. Agencies do not sufficiently promote potential NTFP extraction opportunities (such as wildlife, bushmeat, plants, seeds, fruit, and traditional medicinal plants) in their programs, and provide little guidance to help tribal communities expand beyond subsistence. Due to limited visits, as well as organizational and institutional time constraints, important communications and guidance from external NGOs and the national government often do not reach the tribal communities. As a result, those communities lack sufficient structural guidance on environmental awareness and sustainable resource extraction, which, in turn, aggravates the already severe community shortfalls in essential needs like energy, water, healthcare, and education. The absence of environmental enforcement agencies in or near such communities further reduces basic services, income-generating alternatives, and information, which in turn leads to wildly differing and mostly incorrect interpretations about detrimental activities and how much they should be tolerated by regional communities.

Addressing this breach among all participants—particularly local and national agencies—is and should be a key focus of NGOs. Environmental enforcement for interior communities depends on the extent and quality of the integration of local authority within the national government. Even when they have access to resources and understand the context of subsistence and survival, local areas often do not fully grasp national legislation and fast-changing climate dynamics. Since national institutions typically enjoy far greater awareness of legislation and regulations, they can help all sides learn from each other and strengthen their respective organizations. In the southern region, community-based resource management has already been initiated by the ACT, which to date has been carried out in four communities through the training of 27 Amerindian rangers. But those rangers, even those who participated as field assistants for the Gold Mining Reform Commission (OGS), have no authorization from the national government to enforce existing laws. To make up for that lack of power, eight communities in the southern region signed a declaration to request raising the status of a large section of the area to a protected area (the SSCC) managed by indigenous peoples.

Within the context of REDD+, when Suriname's government has been pressed to hold integrated land use discussions and promote sustainable development on tribal lands in the interior as steps toward the minimization of carbon emission, the momentum for alternative environmental enforcement systems increases. In areas where the encroachment of mining is evident and is condemned by tribal communities, those communities can be further linked with official environmental enforcement bodies to police such activity. Policy discussions on the expansion of hydro-power plants and infrastructural planning have not developed in a linear or participatory manner, for instance, with pressures to address immediate national economic difficulties hijacking long-term needs for sustainable development. And since planning of large-scale projects with potentially large adverse effects on tribal communities is likely to re-emerge, those communities will need to step up awareness-raising and petitioning for environmental and social impact assessments and compensations.

Conclusion

Although Suriname is a country with high biodiversity and a variety of traditional cultures with great potential for community-based resource management and enforcement, it faces multiple challenges at the national, regional, and local levels. Due to in part to legacies of the pre-colonial era, national legislation and regulation of environmental enforcement is dangerously out of sync with the need to increase the authority of local populations. Weak information flow and violations of environmental regulations in the interior also stem from fragmentation of responsibilities among ministries, weak national environmental policy, the absence of a comprehensive umbrella environmental legislation, and limited awareness over the differentiation of rules and regulations between the urban coast and the tribal communities. These conditions underscore the great need for an independent entity to serve as an intermediator between the urban coast and the tribal communities, comprehensively addressing environmental enforcement, managing cultural and language barriers, and advocating community-based sustainable resource management. Based on an understanding of the needs and potential of as well as the challenges and threats faced by Suriname's tribal communities, along with connections within the Surinamese government and international organizations, such an organization could catalyze initiatives and bring parties together to achieve greater balance in the participation of local and non-local authorities in the pursuit of the responsible and transparent development of Suriname's vulnerable rainforest.

Bibliography

Alonso, L.E., and J.H. Mol. 2007. *A Rapid Biological Assessment of the Lely and Nassau plateaus, Suriname (With Additional Information on the Brownsberg Plateau)*. RAP Bulletin of Biological Assessment 43. Arlington, VA: Conservation International.

Colding, Johan, Folke, Carl, (2001) Social Taboos: "Invisible" Systems of Local Resource Management and Biological Conservation. *Ecological Applications* 11 (2): 584

Forest Peoples Program. 2016. *Indigenous peoples in Suriname win important case in the Inter-American Court of Human Rights*. Moreton-in-Marsh, UK: Forest Peoples Program.

Heemskerk, Marieke, and Celine Duijves. 2012. *Child Labor in the Small Scale Gold Mining Industry in Suriname*. Task Order I, Task III. Paramaribo: Research & Data Collection.

Heemskerk, Marieke, Celine Delvoye, and Munjenca Pines. 2015. Interpersonal and Institutional Distrust as Disabling Factors in Natural Resources Management: Small-Scale Gold Miners and the Government in Suriname. *Society and Natural Resources* 28 (2): 133–148.

Ouboter, P., and G.A. Landburg. 2012. Mercury Levels in Pristine and Gold Mining Impacted Aquatic Ecosystems of Suriname, South America. *Ambio* 41: 873–882.

Index

M. Ungar (ed.), *The 21st Century Fight for the Amazon*,
https://doi.org/10.1007/978-3-319-56552-1

CPSIA information can be obtained
at www.ICGtesting.com
Printed in the USA
BVOW05*0213201017
498223BV00012B/32/P

9 783319 565514